Kidney Transplantation: Strategies to Prevent Organ Rejection

Contributions to Nephrology

Vol. 146

Series Editor

Claudio Ronco   *Vicenza*

KARGER

Kidney Transplantation: Strategies to Prevent Organ Rejection

Volume Editors

Claudio Ronco   *Vicenza*
Stefano Chiaramonte   *Vicenza*
Giuseppe Remuzzi   *Bergamo*

5 figures and 13 tables, 2005

Basel · Freiburg · Paris · London · New York ·
Bangalore · Bangkok · Singapore · Tokyo · Sydney

Contributions to Nephrology

(Founded 1975 by Geoffrey M. Berlyne)

....

Claudio Ronco
Department of Nephrology
St. Bortolo Hospital
I-36100 Vicenza (Italy)

Stefano Chiaramonte
Department of Nephrology
St. Bortolo Hospital
I-36100 Vicenza (Italy)

Giuseppe Remuzzi
Department of Medicine and
Transplatation
Ospedali Riuniti Bergamo
I-24128 Bergamo (Italy)

Library of Congress Cataloging-in-Publication Data

Kidney transplantation : strategies to prevent organ rejection / volume
editors, Claudio Ronco, Stefano Chiaramonte, Giuseppe Remuzzi.
    p. ; cm. – (Contributions to nephrology, ISSN 0302-5144 ; v. 146)
  Includes bibliographical references and index.
  ISBN 3-8055-7856-3 (hard cover : alk. paper)
  1. Kidneys–Transplantation–Complications.   2. Graft
rejection–Prevention.   3. Immunosuppresive agents.
  [DNLM: 1. Kidney Transplantation–immunology.   2. Graft
Rejection–immunology.   3. Graft Rejection–prevention & control. WJ 368
K463 2005] I. Ronco, C. (Claudio), 1951- II. Chiaramonte, Stefano. III.
Remuzzi, Giuseppe. IV. Series.
  RD575.K534 2005
  617.4'610592–dc22
                                    2004020573

    Bibliographic Indices. This publication is listed in bibliographic services, including Current Contents® and Index Medicus.

    Drug Dosage. The authors and the publisher have exerted every effort to ensure that drug selection and dosage set forth in this text are in accord with current recommendations and practice at the time of publication. However, in view of ongoing research, changes in government regulations, and the constant flow of information relating to drug therapy and drug reactions, the reader is urged to check the package insert for each drug for any change in indications and dosage and for added warnings and precautions. This is particularly important when the recommended agent is a new and/or infrequently employed drug.

© Copyright 2005 by S. Karger AG, P.O. Box, CH–4009 Basel (Switzerland)
www.karger.com
Printed in Switzerland on acid-free paper by Reinhardt Druck, Basel
ISSN 0302–5144
ISBN 3–8055–7856–3

Contents

V

....

Preface

Transplantation is now firmly established as the therapy of choice for end-stage organ failure. Improvements in surgical techniques and medical management of posttransplant complications, and recent development of novel immunosuppressive strategies has improved the outcome of organ transplantation. Interestingly, this improvement was seen mainly in recipients who never had an acute rejection episode, emphasising the recipients' alloimmune response as a major determinant of overall outcome of the transplant. However, this therapy is not without challenges and risks. Recipients need to continue to take immunosuppressive drugs for the rest of their lives to prevent allograft rejection, and this trades the morbidity and mortality of organ failure for the risks of infection and cancer. In addition, these drugs are likely to contribute to increased mortality from cardiovascular disease, the major cause of premature death in kidney transplant recipients. Moreover, there is the problem of chronic rejection or allograft nephropathy, which arises at least in part because immunosuppressive strategies do not completely inhibit alloimmune response and results in slow progressive deterioration in graft function. These challenges together with the increasing demand of organs for transplantation, create an urgent need for optimizing the outcome of transplantation by achieving long-term, drug-free, graft acceptance with normal organ function. Last year marks the 50th anniversary of Peter Medawar's classic description of acquired immune tolerance, an observation that helped to usher in the modern era of transplantation science and that earned him a share of the 1960 Nobel Prize in Physiology and Medicine. Our understanding of the pathogenesis of transplant rejection and its prevention has progressed enormously in the subsequent years. Since that time, a wealth of experimental data has accumulated relating to strategies for extending allograft survival and function. Recently, numerous insights into the dynamic inter-relationship of host immune responses elicited by donor antigen presentation, either on the graft itself or on specialized antigen-presenting cells have substantially broadened our understanding of the cascade of events that results in the acquisition of tolerance. Still, the question remains of how near we are to the day when long-term tolerance of engrafted organs or tissues is a clinical reality. With the pharmacopoeia of the transplant biologist continually expanding, the potential treatment combinations have become baffling and their impact on strategies to induce tolerance even more complex. It is, therefore, timely to reassess where we stand on the road to achieving clinical transplant tolerance, and highlight the challenges that face us, so that we may choose the best direction in which to invest our efforts in basic and clinical research.

In this 'Course on Kidney Transplantation' a group of world-renowned experts in the field reviews what is new and what is hot in transplantation focussing on novel insights in the pathways of acute rejection and their monitoring through molecular tests, on new immunosuppressive agents in the pipeline, as well as on the most recent and promising approaches to induce tolerance that have emerged from experimental animal studies, with the purpose of understanding whether and how close we are to clinical transplantation.

C. Ronco
S. Chiaramonte
G. Remuzzi

Ronco C, Chiaramonte S, Remuzzi G (eds): Kidney Transplantation: Strategies to Prevent
Organ Rejection. Contrib Nephrol. Basel, Karger, 2005, vol 146, pp 1–10

....

Renal Transplantation

Strategies to Prevent Organ Rejection – The Role of an Inter-Regional Reference Center

Massimo Cardillo[a], Francesca Poli[a], Fiorenza Barraco[a], Nicola De Fazio[a], Giuseppe Rossini[a], Luigi Boschiero[b], Arcangelo Nocera[c], Paolo Rigotti[d], Francesco Marchini[d], Graziella Zacchello[d], Gianfranco Zanon[d], Silvio Sandrini[e], Stefano Chiaramonte[f], Cristina Maresca[g], Rossana Caldara[h], Piergiorgio Messa[i], Luisa Berardinelli[i], Andrea Ambrosini[j], Domenico Montanaro[k], Teresa Rampino[l], Enrico Minetti[m], Eliana Gotti[n], Luciana Ghio[o], Fabrizio Ginevri[p], Fulvio Albertario[q], Mario Scalamogna[a]

[a]Dipartimento Trasfusionale e di Riferimento per il Trapianto di Organi e Tessuti, IRCCS Ospedale Maggiore Policlinico, Milano; [b]Azienda Ospedaliera Borgo Trento, Verona; [c]Azienda Ospedaliera S. Martino, Genova; [d]Azienda Ospedaliera, Padova; [e]Azienda Ospedaliera Spedali Civili, Brescia; [f]Azienda Ospedaliera S. Bortolo, Vicenza; [g]Azienda Ospedaliera Ca' Foncello, Treviso; [h]IRCCS San Raffaele, Milano; [i]IRCCS Ospedale Maggiore Policlinico, Milano; [j]Azienda Ospedaliera Fondazione Macchi, Varese; [k]Azienda Ospedaliera S. Maria della Misericordia, Udine; [l]IRCCS Policlinico S. Matteo, Pavia; [m]Azienda Ospedaliera Ca' Granda Niguarda, Milano; [n]Azienda Ospedaliera Riuniti, Bergamo; [o]Istituti Clinici di Perfezionamento, Milano; [p]IRCCS G. Gaslini, Genova, and [q]Azienda Ospedaliera Istituti Ospitalieri, Cremona, Italy

Abstract

This paper summarizes the role of the Inter-Regional Reference Center (RC) of the North Italy Transplant program (NITp), in coordinating a donor procurement and organ transplantation network, with a special focus on the strategies to minimize immunological risk and complications after transplantation. In the NITp, patients enrolled on the renal transplantation (RT) waiting list are typed for HLA-A,B,DRB1 antigens with a genomic method. They are periodically screened for the presence of lymphocytotoxic antibodies in their serum by the RC and their suitability to receive the transplant is checked periodically. Cadaver kidney allocation is ruled by a computerized algorithm, named NITK3, established in 1997, which aims at ensuring quality, equity, transparency and traceability during all the phases of the allocation decision-making process. NITK3 has been set up by the NITp Working Group on the basis of biological, medical and administrative criteria and it is periodically reviewed after the analysis of transplant results. In this paper, we show the results of a preliminary analysis of RTs performed from 1998 to 2002 in nine out of sixteen centers of the NITp area, which demonstrates the general quality of the NITp program in terms of patients and graft survival and the special attention to the patients at higher immunological risk.

Background

RT is a successful therapy for an ever-increasing number of patients with end-stage renal diseases. Italian data on RTs performed in the years 2000–2001, published by the Centro Nazionale Trapianti (CNT), show a 93% one-year graft survival, higher than that reported by major international registries (UNOS, CNT) [1, 2]. These are the results of an activity, which started more than 30 years ago, with the decisive contribution of the transplant organizations. As a matter of fact, in the 70s clinicians became aware of the importance of HLA matching in RT [3], and of the selection of the most suitable recipient for each donor. It soon became clear that kidney allocation could not be performed by the clinicians responsible for the transplants, but it had to be committed to a super partes facility, able to allocate kidneys according to defined and accepted criteria. In those years, in Italy, Edmondo Malan, Piero Confortini and Girolamo Sirchia founded the NITp, the first transplant organization in Italy. The reference center was set in Milano, where the Blood Transfusion Center of the Ospedale Maggiore Policlinico had already acquired knowledge on leukocyte immunology [4]. On June 18th, 1972, the activity of the NITp began.

The NITp

At present, the NITp is one of the three transplant organizations operating in Italy [5]. It serves an area of about 18 million inhabitants and includes 60 donor-procuring hospitals, 41 transplant centers (16 of whom perform RT), 170 dialysis centers and a single RC in Milano. The tasks of the RC are management of the waiting list, immunological evaluation of recipients and donors, organ allocation, transports organization, data collection, definition of protocols together with the operative units, development of information campaigns, psychological support to donor families and promotion of research and development in the field of organ procurement and transplantation. The NITp serves a defined territory on the basis of official contracts issued by the Regional Health Authorities; this means that patients resident in the NITp regions have free access to the waiting list, whereas nonresidents cannot exceed the 25% of the list in each RT center, which can enroll a maximum of 250 patients. From 1972 to 2003, 5,336 cadaver donors were procured and 9,782 RTs were performed in the NITp area.

HLA Typing, Antibody Screening and Cross-Matching

In the NITp, cadaver donor and recipient HLA-A,B,DR typing is performed using both microcytotoxicity and molecular biology methods [6, 7]. Only broad antigens/alleles are used to assess the degree of donor-recipient mismatching. Sera of the patients enrolled on the RT waiting list are collected every 3 months and 15 days after sensitizing events and screened for the presence of lymphocytotoxic antibodies. Patient medical suitability to receive the transplant is checked at the same time. Panel-reactive antibodies (PRA) are measured against a selected panel of lymphocytes. All the RT recipients must have a negative pre-transplant standard lymphocytotoxicity cross-match against donor T-lymphocytes using the two most recent serum samples. For sensitized patients the current serum plus the two most reactive historical sera are tested.

Renal Allocation

In 1997, the NITp Working Group for Renal Transplantation set up a kidney allocation algorithm from cadaver donors [5], that was named NITK3, to indicate the third algorithm used in the NITp. NITK3 works on the basis of donor and recipients data, entered in the computer by the RC duty officer. Kidney allocation criteria include donor/recipient ABO identity, with a derogation for group B or AB patients at immunological risk (sensitized with PRA equal or above 30% or waiting for a retransplant), who may receive group O or A donor kidneys, respectively, HLA-A,B,DRB1 matching, waiting time on the list, immunological risk, donor and recipient age matching, balance between organs retrieved and RT performed in each center and region. Selected patients are ranked following the above-mentioned biological, medical and administrative variables, and every step of the selection process is registered by the computer in order to guarantee transparency and traceability. NITK3 works as follows: one kidney is offered to the patients belonging to the 'local pool', who are resident in the NITp area and on the waiting list of the RT center in the retrieval zone where the donor has been procured. The first two levels include patients at immunological risk with zero to one and two mismatches with the donor, respectively. The third level includes ordinary patients with zero to one mismatches with the donor and the fourth, ordinary patients with two to four mismatches with the donor. At each level patients are ranked according to the waiting time on the list, that is arranged in three classes (0–3 years, 3–10 years and >10 years), and inside each class patients are listed following the donor-recipient age matching, with a priority for pediatric recipient in the case of a donor below 40 years. If the center responsible for the 'local pool' has no patients with the minimum level of four mismatches with donor, the kidney is allocated to the whole waiting list and a payback is scored. The other kidney is allocated following the rules adopted for the first, but considering the entire waiting list, that includes patients residing outside the NITp area, and a balance is computed among procured and transplanted kidneys. The recipients are proposed to the RT centers strictly following the ranking assuming that all the patients on the list are potentially suitable for transplantation. Pretransplant cross-match is performed following the indications reported in the previous section and must be negative. The NITp duty officer registers all the transactions.

Special Programs

One of the drawbacks in RT is the possibility of successful transplantation in the case of sensitized recipients, as a consequence of previous grafts, pregnancies or transfusions. The antibodies involved in patient sensitization are often HLA-specific and they can be responsible for hyperacute rejection of the graft. For these reasons, sensitized patients or those waiting for a second or third RT are at immunological risk, and the chance of finding a compatible kidney is lower as compared to that of 'non-at-risk' patients. These patients have longer waiting times on the list and for the transplant organizations they are a major problem to deal with. As previously reported, NITK3 kidney allocation algorithm foresees that, among other restrictions, these patients may only receive kidneys with a maximum of two HLA-A,B,DRB1 donor-recipient mismatches. This criterion has been fixed to maximize the chances of a successful RT, but it makes it difficult to find a compatible kidney. A special strategy developed in the NITp for these cases is to identify acceptable donor-recipient mismatches using recently developed laboratory techniques. Donor-recipient acceptable mismatches are defined as donor HLA antigens not present in the recipient, but unable to evoke an immunological response. In the past, the identification of those mismatches by using the lymphocytotoxicity method was very difficult, since it required large HLA-typed cell panels. Moreover this method is ineffective for patients with uncommon phenotypes. Recently, new laboratory methods (Luminex technology [8]) have been developed, which allow the definition of a specific antibody profile for a single patient. Moreover, recent studies have shown that the structure of the HLA antigens can be defined at a molecular level, and only some amino acid triplets in strategic positions on the molecule are presented for the antigen recognition by the host T-cell receptor, thus different HLA antigens in the donor may be unable to be identified as different by the recipient, who does not generate a specific immunological response [9–11]. These assumptions are the basis of a new renal allocation strategy for recipients at immunological risk in the NITp: the number of donor-recipient HLA mismatches is no more an absolute limit for renal allocation, provided that acceptable mismatches and the negativity of pretransplant cross-match are respected. The aim of this program is to improve the transplant chances for the patients at immunological risk without compromising results, and monitoring recipients' post-transplant antibody profile will complete the evaluation. A preliminary experience has been performed in pediatric patients: in the first 3 months, 6 'long waiting' patients at immunological risk have been transplanted. In all cases a good kidney functional recovery was observed. In a period ranging from 1 to 14 days, only one patient showed evidence of vascular rejection, successfully treated with anti-thymocyte globulins plus immunoglobulins.

A second program to be started in the next few months aims at the identification of the role of anti-HLA antibodies in the development of chronic nephropathy after RT. At present, chronic nephropathy, also defined as chronic rejection, is the first cause of RT failure after one year, it is present in about 30–35% of cases after 5 years and it causes a progressive reduction of the glomerular filtration rate. The presence of post-transplant anti-HLA antibodies in the recipient serum is a negative prognostic factor for the development of acute rejection episodes, and it is speculated that they may also play a role in the early phases of immunological events correlated with the vascular injury typical of chronic rejection [12]. The aim of the program is to investigate the post-transplant humoral immunological response against donor HLA antigens and to evaluate the importance of recipient anti-HLA antibodies for the development of late acute rejection episodes and chronic nephropathy after RT.

The NITp RT Registry: Patients, Statistical Methods and Results

As previously reported, data collection is one of the main RC tasks, with the aim of monitoring program quality and giving information to Health Authorities. NITp Operative Units play a crucial role in the registry update, and they are organized as a collaborating network; moreover, in the current year, a safer and faster way to share data will be improved, by the production of a dedicated web that will link together retrieval hospitals, RT centers and the RC. Data quality control completes registry management at the RC. Using the RT data of the NITp Registry, we have performed a statistical analysis to measure the quality of the program in terms of patient and graft survival, and to identify the role of some donor and recipient variables. Patient and graft survival rates were calculated using the actuarial method and the role of the following variables was evaluated by the log rank test: donor and recipient age and donor-recipient gender match, donor-recipient blood group match, donor-recipient body weight ratio, RT center, number of donor-recipient HLA-A,B,DRB1 mismatches, recipient (PRA), number of pretransplant transfusions, waiting time on the list, type of organ utilization, number of previous RTs and cold ischemia time. Patient death was considered as an end point of the graft survival curve regardless of the cause. Information on RT variables, recipient and donor characteristics, graft functional recovery and patient follow-up have been collected in a dedicated database at the RC. The independent contribution of several risk factors on patient and graft survival was evaluated by multivariate analysis (proportional hazard model – PHREG). Values of $p \leq 0.05$ were considered significant. All the statistical analyses were performed using SAS (version 8, SAS Institute, Cary, N.C., USA) statistical package.

One thousand nine hundred and seventy seven RTs from cadaver donors, performed in 1,799 patients from 1998 to 2002 in nine out of sixteen centers of the NITp area have been included in the present study. six hundred and fifty three patients were female (36%), their mean age was 44.8 ± 14.5 years and they were followed up for a mean period of 46.1 ± 19.3 months. Table 1 shows the overall 3-year graft survival and the results of the univariate analysis including the impact of the variables under evaluation. Table 2 shows the results of the multivariate analysis: donor and recipient age are the variables significantly influencing graft survival and, in particular, RT performed with kidneys from donors older than 60 years carry a more than double relative risk of RT failure as compared with those performed with donors aged 19–40 years.

Conclusions

Data reported in the present paper show that RT can be considered one of the greatest successes of the last century's medicine. The preliminary analysis carried out on more than half of the RTs performed in the NITp shows that results are better than those reported by the major international registries. Donor and recipient age are the only variables significantly influencing graft survival in the multivariate analysis, but even when donors or recipients are older than 60 years, 3-year survival rates are above 75%. Though no data are shown on the incidence of graft rejection in our series, no significant difference was demonstrated by univariate and multivariate analysis in 3-year graft survival between patients at immunological risk and 'non-at-risk' patients. Also, these results have

Table 1. Actuarial graft survival

	N	%	3-year graft survival (%)	Standard error (%)	P (log rank)
Overall	1.977	100	86.3	0.8	
Donor age (years)					<0.0001
0–18	214	11	92.5	1.9	
19–40	553	28	91.1	1.3	
41–60	800	40	86.6	1.2	
>60	410	21	75.9	2.2	
Recipient age (years)					<0.0001
0–18	138	7	92.0	2.5	
19–40	543	27	91.7	1.2	
41–60	1.069	54	84.8	1.1	
>60	227	12	77.0	2.9	
Donor-recipient gender match					0.2763
M/M e F/F	1.018	52	87.4	1.1	
M/F	417	21	85.7	1.8	
F/M	542	27	84.7	1.6	
Donor-recipient blood group match					0.2010
Identical	1.937	98	86.2	0.8	
Compatible	40	2	92.5	4.2	
Donor-recipient body weight ratio					0.3354
<0.8	217	11	87.5	2.4	
0.8–1.2	1.076	54	84.8	1.1	
>1.2	684	35	88.3	1.3	
Transplant center					<0.0001
A	340	17	85.1	2.0	
B	301	15	91.0	1.7	
C	262	13	83.1	2.4	
D	249	13	90.9	1.9	
E	238	12	84.8	2.4	
F	188	10	85.6	2.6	
G	182	9	78.4	3.1	
H	100	5	92.4	2.8	
I	60	3	84.4	5.2	
L	57	3	91.0	3.8	
Number of donor-recipient HLA-A,B,DRB1 mismatches					0.3654
0–1	330	17	87.7	1.9	
2–4	1.566	79	86.3	0.9	
5–6	81	4	81.5	4.5	

Table 1 (continued)

	N	%	3-year graft survival (%)	Standard error (%)	P (log rank)
Panel Reactive Antibodies (PRA) (%)					0.8836
0–29	1.805	91	86.5	0.8	
≥30	172	9	84.8	2.9	
Number of pretransplant transfusions					0.1319
0	1.237	63	87.5	1.0	
>0	740	37	84.3	1.4	
Waiting time on the list (years)					0.0566
0–2	966	49	87.1	1.1	
2–5	739	37	86.9	1.3	
>5	272	14	81.8	2.4	
Organ utilization					0.8993
Local	849	43	86.0	1.2	
Shared	1.128	57	86.6	1.1	
Number of previous transplants					0.9001
0	1.799	91	86.4	0.8	
>0	178	9	85.7	2.7	
Cold ischemia time (hours)					0.0788
0–18	1.600	81	87.0	0.9	
>18	377	19	83.3	1.9	

been achieved thanks to the special immunological surveillance on the whole process operated by the RC, from the patient enrollment on the waiting list to the RT surgery and post-transplant follow-up. This surveillance consists in a specific definition of patient immunological profile at the time of enrollment, periodical medical and laboratory screening for sensitizing events, proper organ allocation and constant monitoring of patient survival and rehabilitation. The RC has also the task to assure that the whole transplant process respects the principles of organ availability, equity, transparency, suitability, traceability and efficacy: to this respect, the RC is perfecting a 'Charta of Principles', where these purposes are declared and discussed. One of the main problems to deal with now is the access to RT of some 'penalized' patients, such as those at immunological risk: data from the NITp series under study show that the waiting time before RT for the recipient at immunological risk is 78.8 versus

Table 2. Multivariate analysis: independent contribution of the risk factors on graft survival

	RR	95% CI	p value
Recipient age (years)			
0–18	0.894	0.411–1.943	0.7773
19–40	1.000		
41–60	1.395	0.990–1.965	0.0570
>60	1.878	1.211–2.912	0.0048
Donor age (years)			
0–18	0.811	0.426–1.544	0.5239
19–40	1.000		
41–60	1.439	1.025–2.020	0.0357
>60	2.395	1.642–3.495	<0.0001

30.3 months for 'non-at-risk' recipients. The RC is now considering new patient evaluation and renal allocation strategies, with the aim of finding compatible donor for these recipients and reducing their waiting time on the list. Last, it is to be underlined that every strategy to prevent organ rejection in RT patients, no matter if at immunological risk or not, should be shared among the 'transplant community' providing a coordinated effort by every health worker (immunologist, nephrologist, anesthesiologist, surgeons, coordinators, nurses) who takes part in the transplant process with his specific role.

Acknowledgements

The authors thank for their crucial support the personnel of Donor Procuring Centers, Dialysis Centers, Transplant Centers of the NITp and the organ donor families. A special mention is dedicated to Professors Piero Confortini, Edmondo Malan and Girolamo Sirchia, the founders of the NITp program.

This study was supported in part by the grant 'Progetto di ricerca e di intervento organizzativo a sostegno del processo di donazione e trapianto. Donazione-trapianto: articolazione di una comunicazione complessa' of Centro Nazionale Trapianti.

References

1   Cecka JM: The UNOS Renal Transplant Registry; in Clin Transpl. Los Angeles, UCLA Tissue Typing Laboratory, 2002, pp 1–20.
2   Thomas MA, Luxton G, Moody HR, Woodroffe AJ, Kulkarni H, Lim W, Christiansen FT, Opelz G: Subjective and quantitative assessment of patient fitness for cadaveric kidney transplantation: The 'equity penalty'. Transplantation 2003;75:1026–1029.

3   Bodmer J, Bodmer WF, Payne R, et al: Leucocyte antigens in man: A comparison of lymphocytotoxic and agglutination assays for their detection. Nature 1966;210:28.
4   Sirchia G, Ferrone S, Mercuriali F, et al: Il candidato al trapianto. Aspetti immunologici. Milano, Editrice Ancora, 1969.
5   Sirchia G, Poli F, Cardillo M, Scalamogna M, Rebulla P, Taioli E, Remuzzi G, Nocera A: Cadaver kidney allocation in the North Italy Transplant Program on the eve of the new millennium. Clin Transpl 1998:133–145.
6   Olerup O, Zetterquist H: HLA-DR typing by PCR amplification with sequence-specific primers (PCR-SSP) in 2 hours: An alternative to serological DR typing in clinical practice including donor-recipient matching in cadaveric transplantation. Tissue Antigens 1992;39:225–235.
7   Cesbron-Gautier A, Simon P, Achard L, Cury S, Follea G, Bignon JD: Luminex technology for HLA typing by PCR-SSO and identification of HLA antibody specificities. Ann Biol Clin (Paris) 2004;62:93–98.
8   Pretl K, Chesterton KA, Sholander JT, Leffell MS, Zachary AA: Accurate, rapid characterization of HLA-specific antibody using luminex technology. Hum Immunol 2003;64(suppl 10):S108.
9   Duquesnoy RJ, Howe J, Takemoto S: HLAmatchmaker: A molecularly based algorithm for histo-compatibility determination. IV. An alternative strategy to increase the number of compatible donors for highly sensitized patients. Transplantation 2003;75:889–897.
10   Duquesnoy RJ, Takemoto S, de Lange P, Doxiadis II, Schreuder GM, Persijn GG, Claas FH: HLAmatchmaker: A molecularly based algorithm for histocompatibility determination. III. Effect of matching at the HLA-A,B amino acid triplet level on kidney transplant survival. Transplantation 2003;75:884–889.
11   Dankers MK, Witvliet MD, Roelen DL, de Lange P, Korfage N, Persijn GG, Duquesnoy R, Doxiadis II, Claas FH: The number of amino acid triplet differences between patient and donor is predictive for the antibody reactivity against mismatched human leukocyte antigens. Transplantation 2004;77:1236–1239.
12   Paul LC: Immunologic risk factors for chronic renal allograft dysfunction. Transplantation 2001; 71(suppl 11):S17–S23.

Dr. Massimo Cardillo
Dipartimento Trasfusionale e di Riferimento per il Trapianto di Organi e Tessuti
IRCCS Ospedale Maggiore Policlinico
Via Francesco Sforza 35, IT–20122 Milano (Italy)
Tel. +39 0255034305, Fax +39 0255012573, E-Mail massimo.cardillo@policlinico.mi.it

Ronco C, Chiaramonte S, Remuzzi G (eds): Kidney Transplantation: Strategies to Prevent
Organ Rejection. Contrib Nephrol. Basel, Karger, 2005, vol 146, pp 11–21

··········· ···········

Kidney Transplantation in the Hyperimmunized Patient

James Gloor

Department of Nephrology and Internal Medicine, Mayo Clinic,
Rochester, Minn., USA

Abstract

Individuals who have developed anti-HLA class I and II antibodies are said to be
immunized or sensitized. High levels of donor specific anti-HLA antibodies present at the
time of transplantation frequently result in early allograft loss due to humoral rejection.
Lower levels of donor specific anti-HLA antibodies (DSA) are also associated with poor
outcome. Technological advances in tissue typing permit the detection of low levels of DSA
not seen with standard cytotoxicity cross-match tests. These tests which previously were
used to screen patients to avoid transplantation of donor-immunized patients are now being
used to stratify patients based on their degree of donor alloreactivity. New protocols have
been developed which permit successful transplantation despite the presence of DSA. These
protocols utilize intravenous immunoglobulin infusions prior to transplantation, either alone
or in combination with plasmapheresis to block or remove DSA. Using these protocols many
persons previously considered essentially nontransplantable are now able to successfully
receive transplants. Improved recognition of the clinicopathological characteristics of
humoral rejection have allowed earlier diagnosis and treatment of antibody-mediated allograft injury and improved the outcome. Although these advances have improved the outlook
for highly immunized kidney transplant candidates, more study is needed to delineate the
optimal approach to transplantation in this population.

Definition of the Sensitized Patient

Individuals who have developed anti-HLA antibodies are said to be immunized or sensitized. Typically, these antibodies occur following exposure to nonself HLA antigens, often due to pregnancy, blood transfusion, or previous
transplant [1]. Patients who receive transplants in the presence of donor-specific
anti-HLA antibodies are at risk for the development of humoral rejection, either

immediately upon reperfusion or in the first days to weeks post-transplant [2]. In addition, immunized patients have poorer outcomes compared to nonimmunized individuals even in the absence of donor-specific antibodies (UNOS 2003 annual report www.optn.org/AR2003/509b_can_cur_pra_ki.htm). In an attempt to avoid humoral rejection, techniques have been developed to identify donor specific anti-HLA antibodies (DSA). Initial assays were relatively insensitive, including techniques based on leukoagglutination, and complement-dependent cytotoxicity (CDC). Identification of DSA was considered to contraindicate transplantation. Nevertheless, in some individuals humoral rejection occurred despite negative pretransplant screening for DSA. In an attempt to identify lower levels of DSA, these initial cross-match techniques have been altered in various ways. Incubation periods have been lengthened, additional wash steps have been added, and agents such as anti-human globulin have been incorporated into cross-match assays, with a goal of increasing sensitivity and specificity [3]. In addition to measures taken to increase the sensitivity of cytotoxicity-based cross-match assays, flow cytometric techniques based on antibody adhesion to target cell surface have been developed, and are in use in many transplant centers [4–6]. These assays detect levels of DSA below the level of detection with even sensitivity-enhanced cytotoxicity cross-matches.

In addition to the variability in sensitivity, techniques based on the interaction of the antibody with the target cell membrane are inherently nonspecific, since non-HLA antibodies irrelevant to kidney transplantation may produce a positive cross-match [7, 8]. In an attempt to increase the specificity of anti-HLA antibody detection, solid phase assays have been developed which utilize purified HLA antigens bound to synthetic microparticles or wells [9–12]. These flow- or ELISA-based systems detect HLA antibody activity only, eliminating any non-HLA antigen effect. Importantly, the sensitivity of these antibody detection systems is greater than that seen with complement-dependent cytotoxicity-based assays.

Separate from the level of sensitization against a particular donor, the breadth of anti-HLA antibodies has extremely important implications for transplantation. The panel reactive antibody (PRA) assay is a measure of the number of different anti-HLA antibodies present in an individual at a given timepoint. In this assay, the patient's serum is tested for alloreactivity against a battery of HLA-typed lymphocytes from multiple individuals. The percent of samples exhibiting positive reactions is defined as '% panel reactivity'. Elevated PRA% implies a wide array of anti-HLA antibodies, and makes the likelihood of identification of an acceptable cross-match-negative donor remote. The recognition of this fact has resulted in implementation of a system whereby highly immunized transplant candidates have increased priority with respect to deceased donor organ allocation. This has several ramifications that are not immediately apparent. First, the assay used to determine PRA% is not standardized from one transplant center to another, and the result can vary according to the sensitivity of the cross-match technique used. Alloreactive individuals with PRA determined using less sensitive assays may have lower reported percent positivity, and are thus disadvantaged with respect to individuals studied in centers that use more sensitive assays. Second, the percent PRA is a measure of the breadth of reactivity against a panel of different lymphocytes, but does not measure the specific level of antibody activity against any specific individual. Transplant candidates with very elevated PRA are unlikely to identify a negative cross-match donor. Nevertheless, when the actual antibody activity level against a specific donor is measured by determining the titer of cross-match positivity, frequently this antibody level is found to be very low [13].

Pretransplant Antibody Characterization

Historically, the goal of DSA screening has been detection. In cases in which DSA are identified, that donor/recipient pair has been considered unacceptable, and a different donor sought. The particular antibody detection technique used has been determined by the individual transplant center. Programs that choose highly sensitive cross-match assays such as flow cytometric cross-matching minimize the likelihood of humoral rejection, but some potentially successful transplants may be precluded. Centers using less sensitive assays based on cytotoxicity maximize the number of acceptable transplant pairings, but accept the increased possibility of humoral rejection.

A novel use for cross-match testing is to stratify the degree of immunization. Cross-match assays with different sensitivities and specificities can be combined to provide a detailed assessment of the degree of reactivity between a given donor-recipient pair. High DSA levels are detectable using less-sensitive assays based on complement-dependent cytotoxicity. Lower DSA levels undetected by relatively insensitive assays may be demonstrable with more sensitive techniques such as flow cytometric cross-matching. Solid phase antibody detection systems incorporating purified HLA molecules confirm anti-HLA antibody specificity, including reactivity against specific single HLA antigens, and also exclude positive cell-based cross-match positivity due to irrelevant non-HLA antibodies [14, 15]. This approach enables individuals to be classified according to their level of donor-specific alloreactivity (table 1).

Peritransplant Desensitization

In recent years, techniques have been developed which allow successful kidney transplantation despite the presence of DSA ('positive cross-match kidney

Table 1. Donor-specific alloreactivity based on combining cross-match techniques and antibody detection systems

Antibody level*	Antibody detection technique
High level DSA	Complement-dependent cytotoxicity cross-match positive
Low level DSA	CDC cross-match negative, flow cytometric cross-match positive
Historical DSA	Currently all negative assays, historical serum positive for DSA
Nondonor sensitized	Anti-HLA antibodies detected, but no DSA identified
Nonsensitized	No anti-HLA antibodies detected

*Presence or absence of DSA determined with solid phase antibody detection system.

transplant') [13, 16–19]. These protocols require pretransplant conditioning, either using plasmapheresis (PP) followed by intravenous immunoglobulin (IVIG) or high-dose IVIG alone. Although the optimum technique for abolishing a positive cross-match is not yet determined, both approaches appear to be safe and effective.

High-Dose IVIG

Living Donor Kidney Transplantation. Patients with high levels of DSA based on positive CDC cross-match can be desensitized by the administration of high-dose IVIG, typically 2 g/kg body weight immediately prior to transplantation. Although not completely elucidated, the mechanism by which IVIG abolishes a positive cross-match appears to depend on the interaction of anti-idiotypic antibodies with DSA. This effect is demonstrable within minutes of administration. In addition to the immediate anti-idiotypic effect on cross-match, high-dose IVIG appears to have long-term modulating effects on donor-specific alloreactivity. Patients successfully transplanted using high-dose IVIG desensitization typically maintain negative cross-matches with their donors following transplant. Nevertheless, if this cross-match-negative serum is subjected to interventions to inactivate IgM, the cross-match frequently reverts to positive. This suggests that the synthesis of IgM 'blocking antibodies' with donor specificity plays a role in maintaining nonreactivity following transplant. Other mechanisms have been postulated to explain the effect of high-dose IVIG in preventing humoral rejection in positive cross-match kidney transplants. Among these mechanisms are alterations in cytokine expression or activity, inhibition of complement, induction of B-lymphocyte apoptosis, interference with T-cell activation, and induction of the inhibitory receptor, Fcγ [18]. IVIG may also increase DSA catabolism through its interaction with the IgG transport receptor, FcRn [20]. Although IVIG is frequently effective in neutralizing donor reactivity, some donor/recipient pairs appear unresponsive. Jordan et al. [18] have developed an in vitro assay which successfully predicts IVIG responsiveness. In this assay, recipient serum is incubated with exogenous IVIG, and subsequently subjected to CDC crossmatching with donor T-lymphocytes. In responders, positive cross-match serum exhibits an inhibition of the degree of donor-specific cytotoxicity following incubation with IVIG. Jordan et al. [18] report that 26 of 28 (93%) immunized living donor kidney transplant candidates were characterized as 'IVIG responsive' on the basis of this assay, and all 26 were successfully transplanted, although 2 required repeated doses of IVIG. Immunosuppression consisted of anti-CD25 antibody, tacrolimus, mycophenolate mofetil, and corticosteroids.

Deceased Donor Kidney Transplantation. High-dose IVIG is also an effective tool in deceased donor kidney transplantation. In a prospective, multicenter, placebo-controlled trial comprising twelve transplant centers, serial IVIG versus placebo infusions were administered over 3 years to 101 highly immunized individuals awaiting deceased donor transplants [21]. IVIG significantly lowered PRA and reduced the waiting time to transplantation, compared to the placebo. Twice as many IVIG-treated as placebo-treated patients were transplanted during the study period (37 vs. 17%). In another study, Glotz et al. [16] report that periodic administration of IVIG effectively reduced the PRA in 87% of sensitized patients awaiting transplantation, permitting successful deceased donor transplantation in 11 patients. The mean decrease in PRA in this group was 80%. Post-transplant immunosuppression consisted of thymoglobulin induction, tacrolimus, mycophenolate mofetil, and corticosteroids.

PP/IVIG

Individuals with high levels of DSA can also be transplanted following desensitization with PP followed by low-dose IVIG (100 mg/kg body weight). Serial PP treatments remove DSA, while post PP IVIG replenishes immunoglobulin levels, preventing hypogammaglobulinemia as well as possibly providing an IVIG-mediated immunomodulatory effect. Using this technique, Schweitzer et al. [19] successfully desensitized 11 of 15 positive cross-match kidney transplant candidates. Immunosuppression consisted of OKT3 induction, tacrolimus, mycophenolate mofetil, and corticosteroids. Four of 11 (36%) were diagnosed with acute rejection, but were successfully treated, and patient and allograft survival were 100% after 3–26 months of follow-up. Montgomery et al. [17] used PP/IVIG to successfully transplant 4 patients with positive cross-matches against their living donors. Three of 4 patients had low levels of DSA (CDC cross-match negative, flow cytometric cross-match positive), while one patient had a positive CDC cross-match with the living donor. All 4 developed rejections that were successfully reversed and patient and graft survival were 100% after a mean follow-up of 40 weeks. Notably, the immunosuppression used by this group included induction therapy with the anti-CD25 receptor antibody daclizumab, in addition to tacrolimus, mycophenolate mofetil and corticosteroids. The use of this induction agent permitted them to monitor DSA levels in the post-transplant period. At the time of diagnosis of humoral rejection, DSA levels had increased, and disappeared with further PP/IVIG treatments.

Gloor et al. [13] report a series of 14 anti-human globulin-CDC-positive cross-match kidney transplants. Pretransplant conditioning consisted of PP/IVIG to achieve a negative cross-match at transplantation as well as induction with the anti-CD20 antibody rituximab. Additionally, splenectomy was performed at the time of transplantation. Immunosuppression consisted of thymoglobulin induction, tacrolimus, mycophenolate mofetil, and corticosteroids. Patient and allograft survival were 93 and 79%, respectively, at a mean follow-up of 448 days. Patients were monitored after transplant for humoral rejection with protocol allograft biopsies. Humoral rejection was diagnosed in 44% of patients. In 14% the rejection was clinical (associated with allograft dysfunction), and in 29% it was subclinical.

Low Level DSA

Currently, the optimum technique for positive cross-match transplantation is not clearly defined. Both high-dose IVIG and PP followed by low-dose IVIG protocols are effective and well tolerated in the transplantation of individuals with high levels of DSA. The role of PP may be to lower the circulating DSA to a level below that which results in cellular injury. IVIG may have an immediate, neutralizing effect on DSA-induced cytotoxicity as well as a long term immunomodulatory action on DSA production. Pretransplant conditioning has the goal of lowering circulating DSA activity to below the level causing allograft injury in this group. Typically, this is considered to be a negative CDC cross-match, although DSA may still be detectable using sensitive assays such as flow cytometry, or solid phase assays for HLA antibodies [14, 16, 18].

Individuals with low levels of DSA at baseline may represent a special population. These patients have demonstrable DSA using highly sensitive assays such as flow cytometry, but negative cross-match using less sensitive cytotoxicity-based assays. In this group DSA levels at baseline are already below the goal targeted in published positive cross-match transplant protocols. The function of the desensitization protocol in this population may be to prevent the anamnestic memory response that results in the post-transplant production of high levels of DSA. The risk that low-level DSA poses for patients with positive flow cross-match/negative CDC cross-match is controversial, with some groups reporting poor outcomes, and others reporting no advantage to flow cross-match screening of CDC cross-match negative candidates [4, 22]. Nevertheless, there is an increasing body of evidence that low levels of DSA represent an important risk factor for rejection in the absence of pretransplant conditioning [23, 24]. Preconditioning with high-dose IVIG is effective in preventing humoral rejection in patients with low levels of DSA at baseline (flow cross-match positive/CDC cross-match negative) [25, 26].

Donor Specific Anti-HLA Antibodies following Positive Cross-Match Kidney Transplantation

In the months following a successful positive cross-match kidney transplantation, recipients have negative cross-matches with their donors [13]. Nevertheless, highly sensitive solid phase antibody detection systems identify DSA in the majority of these recipients [14, 27]. The mechanisms by which circulating DSA is maintained at a low level are not clearly defined. Inhibition of DSA production, either by the preconditioning regimen or by the maintenance immunosuppressive regimen may play a role. Nevertheless, third party-specific anti-HLA antibodies are not affected to the same degree as DSA in these patients, implying that mechanisms other than nonspecific suppression of antibody production are involved [14, 27]. The function of the renal allograft itself in maintaining low DSA levels needs to be clarified. Both HLA- and non-HLA-specific antibodies are known to be bound by renal allografts, raising the possibility that antibody trapping and catabolism may occur [28].

Accommodation is the term that has been applied to the condition in which donor-specific antibody coexists with its target antigen without resultant allograft injury [29]. Although this phenomenon has been described predominantly in ABO incompatible and xenotransplantation, there is evidence that exposure of renal allograft endothelium to low levels of anti-HLA antibodies results in the up-regulation of expression of protective genes such as Bcl-xL [30]. Using microarray analysis, Park et al. [31] found significant differences in intragraft gene expression in ABO incompatible and compatible kidney transplants, providing evidence that accommodation is a process occurring at the level of the allograft. Whether similar intragraft changes in gene expression occur in response to exposure to anti-HLA antibodies with donor reactivity remains to be determined.

Humoral Rejection

The clarification of the distinct clinical and histological characteristics of humoral rejection represents an important advance in the management of immunized kidney transplant recipients. Multiple factors have contributed to this improved characterization. Increasing numbers of kidney transplants performed across humoral barriers such as the ABO blood group as well as positive cross-match transplantation has increased the frequency of humoral rejection. Antibody-mediated injury is increasingly considered in the differential diagnosis of allograft dysfunction. Technological advances in histocompatibility testing have significantly improved the ability of transplant centers to identify DSA. More individuals who are alloimmunized from prior transplants are being retransplanted. Finally, deposition of the complement degradation product C4d, a component of the classical complement pathway, in peritubular capillaries has been identified as a sensitive and specific marker for humoral rejection [32, 33].

Although humoral and acute cellular rejections frequently occur together, there is a relatively specific constellation of clinical and histological findings typical of antibody-mediated allograft injury. Trpkov et al. [34] identified severe vasculitis, glomerulitis, fibrin thrombi in glomeruli, neutrophilic infiltration of peritubular capillaries, and fibrinoid necrosis as being more commonly detected in acute rejection when DSA are identifiable. Mononuclear cell interstitial infiltrate and tubulitis typical of acute cellular rejection is not a feature commonly associated with isolated antibody-mediated rejection. Clinical features associated with acute rejection episodes with associated DSA compared to non-DSA rejections include poorer function and lower allograft survival 6 months after transplant, refractoriness to anti-lymphocyte therapy, and higher incidence of steroid resistance [35]. The recognition of the distinct clinical and morphological characteristics of humoral rejection has led to a modification of the Banff 1997 classification of renal allograft rejection [36]. In this system three criteria are included in the diagnosis of antibody-mediated rejection. These criteria include (1) morphological evidence of acute tissue injury such as acute tubular necrosis, neutrophils and/or mononuclear cells or thrombi in peritubular capillaries or glomeruli, or intimal arteritis, fibrinoid necrosis, or intramural inflammation in arteries; (2) immunopathological evidence for antibody action such as C4d deposition in peritubular capillaries, or immunoglobulin and complement deposition in areas of arterial fibrinoid necrosis, and (3) serological evidence of circulating antibodies to donor HLA or other anti-donor endothelial antigens. In the absence of identifiable DSA, or peritubular capillary C4d deposition, light microscopic findings consistent with humoral rejection are reported as 'suspicious' for antibody-mediated rejection.

Treatment of Humoral Rejection

The diagnosis of humoral rejection is important for more than just theoretical reasons. Humoral rejection is frequently unresponsive to measures directed toward acute cellular rejection [35]. Similar to pretransplant conditioning regimens for positive cross-match kidney transplantation, high-dose IVIG and PP followed by low-dose IVIG have been reported to be efficacious in the therapy of humoral rejection [13, 17, 35, 37].

Conclusion

Alloimmunization represents a significant barrier to kidney transplantation. Nevertheless, important technical advances in antibody screening and cross-matching permit much more comprehensive characterization of DSA, both with respect to sensitivity and specificity. New patient management protocols have been developed which permit kidney transplantation across a positive cross-match. Finally, new histopathological approaches permit more accurate diagnosis of humoral rejection, and therapeutic interventions are available to reverse it. These new developments have greatly improved the prospects for successful transplantation in immunized individuals with end-stage renal disease.

References

1   Moore SB, Sterioff S, Pierides AM, Watts SK, Ruud CM: Transfusion-induced alloimmunization in patients awaiting renal allografts. Vox Sanguinis 1984;47:354–361.
2   Kissmeyer-Nielsen F, Olsen S, Petersen VF, Fjelborg O: Hyperacute rejection of kidney allografts, associated with pre-existing humoral antibodies against donor cells. Lancet 1966;2:662–665.
3   Zachary AA, Klingman L, Thorne N, Smerglia AR, Teresi GA: Variations of the lymphocytotoxicity test. An evaluation of sensitivity and specificity. Transplantation 1995;60:498–503.
4   Mahoney RJ, Norman DJ, Colombe BW, Garovey MR, Leeber DA: Identification of high- and low-risk second kidney grafts. Transplantation 1996;61:1349–1355.
5   Berteli AJ, Daniel V, Mohring K, Staehler G, Opelz G: Association of kidney graft failure with a positive flow cytometric crossmatch. Clin Transplant 1992;6:31–34.
6   Scornik JC, Brunson ME, Schaub B, Howard RJ, Pfaff WW: The crossmatch in renal transplantation. Transplantation 1994;57:621–625.

7   Ten Hoor GM, Coopmans M, Allebes WA: Specificity and Ig class of preformed alloantibodies causing a positive crossmatch in renal transplantation. Transplantation 1993;56:298–304.

8   Lobashevsky A, Senkbeil R, Shoaf J, Rowe C, Lobashevsky E, Burke R, Hudson SL, Deierhoi MH, Thomas J: Specificity of preformed alloantibodies causing B cell positive flow crossmatch in renal transplantation. Clin Transplant 2000;14:533–542.

9   Rebibou JM, Chabod J, Bittencourt MC, Threvenin C, Chalopin JM, Herve P, Tiberghien P: Flow-PRA evaluation for antibody screening in patients awaiting kidney transplantation. Transpl Immunol 2000;8:125–128.

10   Pei R, Lee JH, Shih NJ, Chen M, Terasaki PI: Single human leukocyte antigen flow cytometry beads for accurate identification of human leukocyte antigen antibody specificities. Transplantation 2003;75:43–49.

11   Lucas DP, Paparounis ML, Myers L, Hart JM, Zachary AA: Detection of HLA class I specific antibodies by the QuikScreen enzyme-linked immunosorbent assay. Clin Diagn Lab Immunol 1997;(May):252–257.

12   Zachary AA, Delaney NL, Lucas DP, Lefell MS: Characterization of HLA class I specific anti-bodies by ELISA using solubilized antigen targets: I Evaluation of GTI QuikID assay and analy-sis of antibody patterns. Hum Immunol 2001;62:228–235.

13   Gloor JM, DeGoey SR, Pineda AA, Moore SB, Prieto ML, NS, Larson TS, Griffin MD, Textor SC, Velosa JA, Schwab TR, Fix LA, Stegall MD: Overcoming a positive crossmatch in living donor kidney transplantation. Am J Transplant 2003;3:1017.

14   Gloor JM, DeGoey S, Ploeger N, Gebel HM, Bray RA, Moore SB, Dean PG, Stegall MD: Persistence of low levels of alloantibody after desensitization in crossmatch positive living donor kidney transplantation. Transplantation 2003;in press.

15   Gebel HM, Bray RA, Nickerson P: Pre-transplant assessment of donor-reactive HLA-specific antibodies in renal transplantation: Contraindication vs. risk. Am J Transplant 2003;in press.

16   Glotz D, Antoine C, Julia P, Suberbielle-Boissel C, Boudjeltia S, Fraoui R, Hacen C, Duboust A, Bariety J: Desensitization and subsequent kidney transplantation of patients using intravenous immunoglobulin. Am J Transplant 2002;2:758–760.

17   Montgomery RA, Zachary AA, Racusen LC, Leffell MS, King KE, Burdick J, Maley WR, Ratner LE: Plasmapheresis and intravenous immune globulin provides effective rescue therapy for refractory humoral rejection and allows kidneys to be successfully transplanted into cross-match-positive recipients. Transplantation 2000;70:887–895.

18   Jordan SC, Vo A, Bunnapradist S, Toyoda M, Peng A, Puliyanda D, Kamil E, Tyan D: Intravenous immune globulin treatment inhibits crossmatch positivity and allows for successful transplantation of incompatible organs in living-donor and cadaver recipients. Transplantation 2003;76:631–636.

19   Schweitzer E, Wilson JS, Fernandez-Vina M, Fox M, Gutierrez M, Wiland A, Hunter J, Farney A, Philosophe B, Colonna J, Jarrell BE, Bartlett ST: A high panel-reactive antibody rescue protocol for cross-match-positive live donor kidney transplants. Transplantation 2000;70:1531–1536.

20   Yu Z, Lennon VA: Mechanism of intravenous immune globulin therapy in antibody-mediated autoimmune diseases. N Engl J Med 1999;340:227–228.

21   Jordan SC, Tyan D, Stablein DM, McIntosh M, Vo A: Evaluation of intravenous immunoglob-ulin (IVIG) as an agent to lower allosensitization and improve transplantation in highly-sensitized adult ESRD patients: Report of the NIH IG02 trial. Am J Transplant 2003; 3(supp 5):551.

22   Christiaans MHL, Overhop R, Ten Haaft A, Nieman F, van Hooff JP, van den Berg-Loonen EM: No advantage of flow cytometry crossmatch over complement-dependent cytotoxicity in immunologically well-documented renal allograft recipients. Transplantation 1996;62: 1341–1347.

23   Karpinski M, Rush D, Jeffery J, Exner M, Regele H, Dancea S, Pochinco D, Birk P, Nickerson P: Flow cytometric crossmatching in primary renal transplant recipients with a negative anti-human globulin enhanced cytotoxicity crossmatch. J Am Soc Nephrol 2001;12:2807–2814.

24   Nickerson P, Karpinski M, Gibson IW, Jeffery J, Rush D: HLA DR mismatching and donor spe-cific antibodies are risk factors for acute rejection even in the era of modern immunosuppression. Am J Transplant 2004;4(supp 8):258.

25   Akalin E, Ames S, Sehgal V, Fotino M, Daly L, Murphy B, Bromberg JS: Intravenous immunoglobulin and thymoglobulin facilitate kidney transplantation in complement-dependent cytotoxicity B-cell and flow cytometry T- or B-cell crossmatch positive patients. Transplantation 2003;76:1444–1447.

26   Gloor JM, Mai ML, DeGoey S, Larson TS, Gonwa TA, Gebel HM, Bray RA, Moore SB, Genco PV, Stegall MD: Kidney transplantation following administration of high dose intra-venous immunoglobulin in patients with positive flow cytometric/negative enhanced cytotox-icity crossmatch. Am J Transplant 2004;4(supp 8):256.

27   Zachary AA, Montgomery RA, Ratner LE, Samaniego-Picota M, Haas M, Kopchaliiska D, Lefell MS: Specific and durable elimination of antibody to donor HLA antigens in renal transplant patients. Transplantation 2003;76:1519–1525.

28   Joyce S, Flye MW, Mohankumar T: Characterization of kidney-cell specific, non-major histo-compatability complex alloantigen using antibodies eluted from rejected human renal allografts. Transplantation 1988;46:362–369.

29   Platt JL, Vercellotti GM, Dalmasso AP, Matas AJ, Bolman RM, Najarian JS, Bach FH: Transplantation of discordant xenografts: A review of progress. Immunol Today 1990;11:450–456.

30   Salama AD, Delikouras A, Pusey CD, Cook HT, Bhangal G, Lechler RI, Dorling A: Transplant accomodation in highly sensitized patients: A potential role for Bcl-xL and alloantibody. Am J Transplant 2001;1:260–269.

31   Park WD, Grande JP, Ninova D, Nath KA, Platt JL, Gloor JM, Stegall MD: Accommodation in ABO-incompatible kidney allografts. Am J Transplant 2003;3:952.

32   Feucht HE, Schneeberger H, Hillebrand G, Burkhardt K, Weiss M, Riethmuller G, Land W, Albert E: Capillary deposition of C4d complement fragment and early renal graft loss. Kidney Int 1993;43:1333–1338.

33   Collins AB, Schneeberger EE, Pascual M, Saidman SL, Williams WW, Tolkoff-Rubin N, Cosimi AB, Colvin RB: Complement activation in acute humoral renal allograft rejection: Diagnostic signifi-cance of C4d deposits in peritubular capillaries. J Am Soc Nephrol 1999;10:2208–2214.

34   Trpkov K, Campbell P, Pazderka F, Cockfield S, Solez K, Halloran PF: Pathologic features of acute renal allograft rejection associated with donor-specific antibody. Analysis using the Banff Grading Schema. Transplantation 1996;61:1586–1592.

35   Crespo M, Pascual M, Tolkodd-Rubin N, Mauiyyedi S, Collins BA, Fitzpatrick D, Linn Farrell M, Williams WW, Delmonicao FL, Cosimi BA, Colvin RB, Saidman SL: Acute humoral rejection in renal allograft recipients: I. Incidence, serology, and clinical characteristics. Transplantation 2001;71:652–658.

36   Racusen LC, Colvin RB, Solez K, Mihatsch MJ, Halloran PF, Campbell PM, Cecka MJ, Cosyns J-P, Demetris AJ, Fishbein MC, Fogo A, Furness P, Gibson IW, Glotz D, Hayry P, Hunsickern L, Kashgarian M, Kerman R, Magil AJ, Montgomery R, Morozumi K, Nickeleit V, Randhawa P, Regele H, Seron D, Seshan S, Sund S, Trpkov K: Antibody-mediated rejection criteria – an addition to the Banff '97 classification of renal allograft rejection. Am J Transplant 2003;3:708–714.

37   Jordan SC, Quartel AW, Czer LSC, Admon D, Chen G, Fishbein MC, Schweiger J, Steiner RW, Davis C, Tyan DB: Posttransplant therapy using high dose human immunoglobulin (intravenous gammaglobulin) to control acute humoral rejection in renal and cardiac allograft recipients and potential mechanism of action. Transplantation 1998;66:800–805.

James Gloor, MD
Department of Nephrology and Internal Medicine
Mayo Clinic, 200 1st St SW
Rochester, MN 55905 (USA)
E-Mail gloor.james@mayo.edu

Ronco C, Chiaramonte S, Remuzzi G (eds): Kidney Transplantation: Strategies to Prevent
Organ Rejection. Contrib Nephrol. Basel, Karger, 2005, vol 146, pp 22–29

....

Chronic Induction

What's New in the Pipeline

Flavio Vincenti

University of California, San Francisco, Calif., USA

Abstract

Induction therapy with biological agents was introduced the in the 1970s and the
rationale, concepts and approach have remained almost unchanged for 30 years. However,
the novel biological agents being developed for induction therapy are being designed for
chronic rather than short-term therapy with several objectives: reduce dependence on toxic
and nephrotoxic agents, improve outcome and ultimately facilitate the emergence of tolerance. The biological agents include efalizumab, a humanized anti-CD11a (anti-LFA1), anti-
CD154, anti-CD40, a number of agents targeting IL-15 and its receptor, and costimulation
blockade with humanized antibodies to CD80/CD86 and the fusion receptor protein
LEA29Y, a second generation $CTLA_4Ig$. The past decade has witnessed an unprecedented
number of small molecules/oral drugs that have been developed and approved for renal
transplantation; the next decade, however, may witness the emergence of a new class of
biological induction agents that may displace some of the currently used drugs.

Introduction

Historically induction therapy was conceived for the modification of the
host immune system at the time of antigen presentation. The purpose of induction
therapy with biological agents was to reorient the immune system by depleting
potentially alloreactive immune cells. While not conclusively proven, the eventual
re-emergence of re-educated T-cells was hypothesized to lead to graft tolerance
(i.e., graft acceptance) rather than rejection. This led to the use of a number of
agents such as the polyclonal anti-lymphocytes and the murine anti-CD3 OKT3
[1, 2]. The introduction of the anti-interleukin-2 receptor antibodies (anti-IL-2R
mAbs) with their long life, lack of side effects and prolonged biological effects
have not, however, resulted in a sustained effort to use these new biologics for

chronic induction therapy [3, 4]. The newest entry, Campath 1H, a humanized anti-CD52 mAb has been tested predominately in investigator-initiated nonrandomized trials [5, 6]. While the function of CD52 is unclear, its expression on T and B cells results in the prolonged depletion after administration of one to two doses of Campath 1H. This is a unique example of short-term induction therapy with prolonged biological activity.

The pipeline for biological agents, however, is focused on chronic induction therapy rather than immunosuppression coverage with a brief intervention regimen. There are several reasons for this approach. The first is that the mechanism of action of the new agents make them more effective yet safe with prolonged therapy. In addition, these novel biological agents are being utilized in immunosuppression regimens that eliminate calcineurin inhibitors (CNIs) and/or corticosteroids thus necessitating an extended period of therapy. Another important reason for this approach is economic. The cost of clinical development of a new biologic for transplantation can be justified only if it is designed for chronic and prolonged (possibly indefinite) use. Two possible exceptions for the use of biologics long-term include the biological agents being considered to prevent the ischemia-reperfusion injury associated with delayed graft function. The other is the use of anti-CD20 to deplete B cells to prevent humoral rejection or/and interfere with antigen presentation at the time of transplantation.

Novel Agents for Chronic Induction

The novel agents being developed for chronic induction and repeated use offer advantages for immunosuppression (table 1). The new agents, whether chimeric, humanized monoclonal antibodies or fusion receptor proteins, have long half-life and prolonged biological effects. They lack immunogenicity and can be reused chronically. Their administration is not associated with acute toxicities or cytokine release. The mechanism of action of these novel agents as well as their use in protolerogeneic immunosuppression regimens can lead to better graft acceptance and possibly tolerance. Table 2 lists the current biological agents and their status in clinical development.

Efalizumab

Efalizumab is a humanized IgG1 monoclonal antibody targeting the CD11a chain of LFA1. Efalizumab binds to LFA1 preventing LFA1-ICAM interaction. Anti-CD11α has been shown to block T-cell adhesion, trafficking and activation [7]. Pretransplant therapy with anti-CD11a prolonged survival of

Table 1. Potential advantages of chronic induction therapy

- Antibodies and fusion-receptor proteins have long half-life and prolonged biological effects
- Lack immunogenicity
- Recurrent intermittent use
- Lack acute toxicities
- Targets can be readily saturated
- Pro 'tolerance' effects
- Can displace use of calcineurin inhibitors and steroids

Table 2. Novel biologics for chronic induction

Agent	Pharma or Biotech	Status
Anti-IL15/mIL15 Fcγ	Amgen-Roche	Preclinical
Anti-CD154	Biogen-IDEC-Novartis	On hold
Anti-CD40	Bristol Myers-Chiron	Preclinical
Efalizumab (anti-CD11a)	Xoma, Genetech	Phase II
h1F1 and h3D1 (anti-CD80/CD86)	Wyeth	Phase I-on hold
LEA29Y (second generation CTLA4Ig)	Bristol Myers	Phase II → III

murine skin and heart allografts and monkey heart allografts [8]. Efalizumab has been approved for use in patients with psoriasis. In a phase I/II open label, dose ranging, multidose, multicenter trial, Efalizumab was administered subcutaneously, weekly for 12 weeks following renal transplantation [9]. Table 3 shows the doses of efalizumab that were used in this study. Efalizumab was used as chronic induction (for 3 months) with a maintenance regimen of full dose or half dose of cyclosporine. At 3 months 7.8% of patients had reversible rejection episodes and at 6 months there was one additional rejection for a cumulative rejection rate of 10.4%. Pharmacokinetic and pharmacodynamic studies showed that the lower doses of Efalizumab (0.5 mg/kg) produced saturation and 80% modulation of CD11a within 24 h of therapy. In a subset of 10 patients who received the high dose Efalizumab (2 mg/kg) with full dose cyclosporine, mycophenolate mofetil (MMF) and steroids, 3 of 10 patients developed post-transplant lymphoproliferative diseases. While Efalizumab appears to be an effective immunosuppressive agent, it should be used in a lower dose (0.5 mg/kg) and with an immunosuppressive regimen that spares calcineurin inhibitors.

Table 3. Efalizumab dose and concomitant immunosuppression

Dose of Efalizumab	Group I 0.5 mg/kg	Group II 2.0 mg/kg
A: 1/2 dose CsA + sirolimus + prednisone	n = 9	n = 9
B: Full dose CsA + MMF + prednisone	n = 10	n = 10
Totals (n = 38)	n = 19	n = 19

CsA = Cyclosporine; MMF = mycophenolate mofetil.

Anti-IL-15 Therapy

A multicenter study with anti-IL-2 mAbs induction and MMF with steroids (CNIs free) resulted in an incidence of acute rejection of 48% at 6 months [10]. The rejection was probably mediated by IL-15 as both circulating and intragraft lymhocytes had fully saturated IL-2 receptor with the anti-IL-2 mAbs. IL-15 has been shown to mediate the escape rejection during therapy with anti-IL-2 mAbs [11, 12]. Anti-IL-15 mAbs, anti-IL-15 α receptor mAbs and a fusion receptor mutant IL-15 may be developed for chronic induction for use with the anti-IL-2 mAbs in a CNI free regimen [13].

Costimulation Blockade

The CD154-CD40 pathway originally described in the activation of B-cells, is also important in T cell activation. Following the impressive results of anti-CD154 in nonhuman primates, a phase I study of Hu5C8 (humanized anti-CD154) was initiated with chronic intermittent administration of Hu5C8 (short course of steroid, but CNI free) [14, 15]. Unfortunately, the study was halted following the occurrence of several thromboembolic complications [14].

Another humanized anti-CD154, IDEC 131, IDEC Pharmaceuticals was in clinical trials in patients with autoimmune diseases. However, recent thromboembolic events in several patients treated with IDEC 131 raised the concern that antibodies that bind to CD154 may not be safe. This study has been put on hold.

In fact, yet another mAb to CD154 from Novartis was also recently found to result in thromboembolism in nonhuman primates [16]. While the CD154-CD40 remains a promising therapeutic target, it is possible that blockade of the CD40 receptor may prove a safer approach.

The most characterized costimulatory pathway is CD28-CD80/CD86 [17]. T-cell activation requires two signals: the first is delivered by allopeptides to the T-cell receptor (defining the specificity of the response) and the second is mediated by CD28 following its ligation by CD80/CD86. Without costimulation, the T-cell does not proliferate, does not produce cytokines, becomes anergic and undergoes apoptosis. Several experimental models have confirmed the potential of costimulation blockade in inducing either tolerance or effective immunosuppression [18, 19]. Two approaches have been used to block CD28-mediated T-cell activation. The first was to target CD80/CD86 with the humanized monoclonal antibodies h1F1 and h3D1 [20]. In vitro h1F1 and h3D1 were shown to block CD28-dependent T-cell proliferation and decrease mixed lymphocyte reactions. The monoclonal antibodies need to be used in tandem since neither CD80 nor CD86 are sufficient to stimulate T-cells via CD28. Anti CD80 and CD86 mAbs were shown to be effective in renal transplantation in nonhuman primates, in monotherapy or in combination with steroids or cyclosporine [18, 21]. Their use, however, did not result in durable tolerance. A phase I study of h1F1 and h3D1 in renal transplant recipients was performed in patients receiving maintenance therapy consisting of cyclosporine, MMF and steroids [20]. Patients received a single pretransplant dose ranging from 1.5 to 5 mg/kg of each mAb.

The preliminary results of this phase I study in 24 patients showed that h1F1 and h3D1 were safe, but additional studies with chronic therapy are required to determine their efficacy.

The second approach to block the CD28-CD80/CD86 pathway is to use the fusion receptor protein CTLA4Ig (the extracellular portion of CTLA4 fused to the Fc fragment of IgG1). CTLA4 is homologous to CD28, is up-regulated after T-cell activation, has higher avidity to CD80/CD86 than CD28. However, unlike CD28, CTLA4 transduces negative signals and interrupts T-cell activation [17]. CTLA4 sequesters CD80/CD86 and blocks their binding to CD28. LEA29Y is a second generation CTLA4Ig which has been reengineered with two point mutations in the CTLA4 binding sites to increase the avidity to CD80 (2-fold) and CD86 (4-fold). 21 LEA29Y is 10-fold more effective than CTLA4Ig in vitro on a per dose basis in inhibiting T-cell effector functions [22].

A phase III trial of chronic LEA29Y therapy with MMF and steroids in a CNI-free regimen was recently presented at the American Transplant Congress 2004 [22]. The immunosuppression protocol and the LEA29Y regimen are shown in figure 1. Two-hundred and seventeen primary or retransplants were randomized to receive one of two regimens of LEA29Y or a cyclosporine-based immunosuppression therapy. At 6 months, acute rejection was similar between the LEA29Y-treated patients and the cyclosporine-treated patients [22]. Therapy

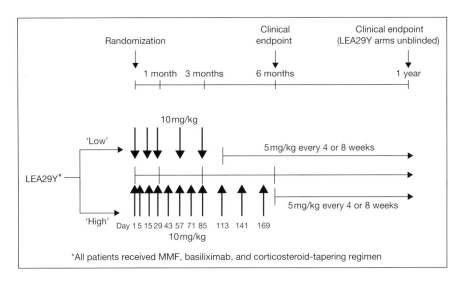

Fig. 1. LEA29Y phase II dose-finding study design.

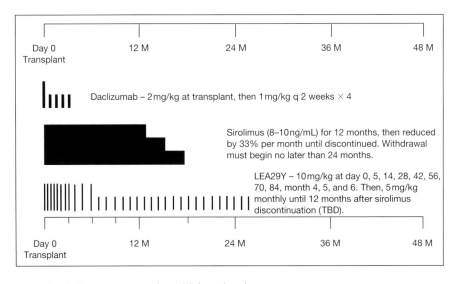

Fig. 2. Immunosuppressive withdrawal regimen.

with LEA29Y was associated with better renal function, lower blood pressure levels and lower LDL values when compared to cyclosporine-treated patients [22]. This study demonstrated that chronic induction therapy of LEA29Y can replace cyclosporine [23].

Ultimately for chronic induction therapy to be widely accepted it may have to facilitate the elimination of both CNIs and steroids and deliver improved outcome (i.e., less toxicity) with immunosuppression drug simplification.

A prototype of this approach is a primate study reported by Adams et al. [24]. In this trial pancreatectomized nonhuman primates had prolonged allogeneic islet cells survival with therapy with LEA29Y and sirolimus. Based on this study, we plan to conduct a trial (in collaboration with Dr. Chris Larsen at Emory University) supported by the Immune Tolerance Network in recipients of living donor kidneys with chronic induction with LEA29Y (and a brief anti-IL-2 mAb induction) and sirolimus monotherapy as shown in figure 2. Patients with no evidence of rejection and a quiescent anti-donor immunological profile may be withdrawn of sirolimus at one year and LEA29Y at 2 years after transplantation. This immunosuppression regimen allows complete withdrawal in selected patients that exhibit tolerance to the allograft.

In summary, chronic induction therapy offers a wide range of therapeutic opportunities to decrease dependence on toxic drugs, improve outcome, allow drug minimization and simplification and ultimately facilitate induction of tolerance.

References

1   Szczech LA, Berlin JA, Aradhye S, Grossman RA, Feldman HI: Effect of anti-lymphocyte induction therapy on renal allograft survival: A meta-analysis. J Am Soc Nephrol 1997;8:1771–1777.
2   Ortho Multicenter Transplant Study Group. A randomized clinical trial of OKT3® monoclonal antibody for acute rejection of cadaveric renal transplants. N Engl J Med 1985;313:337–342.
3   Vincenti F, Kirkman R, Light S, Bumgardner G, Pescovitz M, Halloran P, Neylan J, Wilkinson A, Ekberg H, Gaston R, Backman L, Burdick J: Interleukin 2 receptor blockade with daclizumab to prevent acute rejection in renal transplantation. N Eng J Med 1998;338:161–165.
4   Nashan B, Moore R, Amlot P, Schmidt AG, Abeywickrama K, Soulillou J: Randomised trial of basiliximab versus placebo for control of acute cellular rejection in renal allograft recipients. Lancet 1997;350:1193–1198.
5   Kirk AD, Hale DA, Mannon RB, Kleiner DE, Hoffmann SC, Kampen RL, Cendales LK, Tadaki DK, Harlan DM, Swanson SJ: Results from a human renal allograft tolerance trial evaluating the humanized CD52-specific monoclonal antibody alemtuzumab (CAMPATH-1H). Transplantation 2003;76:120–129.
6   Knechtle SJ, Pirsch JD, H Fechner J Jr, Becker BN, Friedl A, Colvin RB, Lebeck LK, Chin LT, Becker YT, Odorico JS, D'Alessandro AM, Kalayoglu M, Hamawy MM, Hu H, Bloom DD, Sollinger HW: Campath-1H induction plus Rapamycin monotherapy for renal transplantation: Results of a pilot study. Am J Transplant 2003;3:722–730.
7   Nakakura EK, Shorthouse RA, Zheng B, McCabe SM, Jardieu PM, Morris RE: Long-term survival of solid organ allografts by brief anti-lymphocyte function-associated antigen-1 monoclonal antibody monotherapy. Transplantation 1996;62:547–552.
8   Salmela K, Wramner L, Ekberg H, Hauser I, Bentdal O, Lins L-E, Isoniemi H, Backman L, Persson N, Neumayer H-H, Jorgensen PF, Spieker C, Hendry B, Nicholls A, Kirste G, Hasche G: A randomized multicenter trial of the anti-ICAM-1 monoclonal antibody (enlimomab) for the prevention of acute rejection and delayed onset of graft function in cadaveric renal transplantation. Transplantation 1999;67:729–736.

9   Vincenti F, Mendez R, Rajagopalan PR, et al: A phase I/II trial of anti-CD11a monoclonal antibody in renal transplantation. Am J Transplant 2001;1(suppl 1):276(abstract).

10   Vincenti F, Ramos E, Brattstrom C, Cho S, Ekberg H, Grinyo J, Johnson R, Kuypers D, Stuart F, Khanna A, Navarro M, Nashan B: Multicenter trial exploring calcineurin inhibitors avoidance in renal transplantation. Transplantation 2001;71:1282–1287.

11   Pavlakis M, Strehlau J, Lipman M, Shapiro M, Maslinski W, Strom TB: Intragraft IL-15 transcripts are increased in human renal allograft rejection. Transplantation 1996;62:543–545.

12   Baan CC, Knoop CJ, van Gelder T, Holweg CTJ, Niesters HGM, Smeets TJM, van der Ham F, Zondervan PE, Maat LPWM, Balk AHMM, Weimar W: Anti-CD25 therapy reveals the redundancy of the intragraft cytokine network after clinical heart transplantation. Transplantation 1999;67: 870–876.

13   Zheng XX, Sanchez-Fueyo A, Masayuki S, Dmenig C, Sayegh MH, Strom TB: Favorable tipping the balance between cytopathic and regulatory cells to create transplantation tolerance. Immunity 2003;19:503–514.

14   Kirk AD, Knechtle SJ, Sollinger HW, Vincenti FG, Stecher S, Nadeau K: Preliminary results of the use of humanized anti-CD154 in human renal allotransplantation (abstract #223). Am J Transplant 2001;1(suppl 1):190.

15   Kirk AD, Burkly LC, Batty DS, Baumgartner RE, Berning JD, Buchanan K, Fechner JH Jr, Germond RL, Kampen RL, Patterson NB, Swanson SJ, Tadaki DK, TenHoor CN, White L, Knechtle SJ, Harlan DM: Humanized anti-CD154 monoclonal antibody treatment prevents acute renal allograft rejection in non-human primates. Nat Med 1999;5:686.

16   Kanmaz T, Fechner J, Torrealba J, Kim H, Dong Y, Oberley T, Schultz J, Schuler W, Hu H: Monotherapy with the human anti-CD154 monoclonal antibody AB1793 prolongs allograft survival in rhesus monkeys. Am J Transplant 2003;3:A599.

17   Sayegh MH, Turka LA: The role of T-cell costimulatory activation pathways in transplant rejection. N Engl J Med 1998;448:1813–1821.

18   Kirk AD, Tadaki DK, Celniker A, Batty DS, Berning JD, Colonna JO, Cruzata F, Elster EA, Gray GS, Kampen RL, Patterson NB, Szklut P, Swanson J, Xu H, Harlan DM: Induction therapy with monoclonal antibodies specific for CD80 and CD86 delays the onset of acute renal allograft rejection in non-human primates. Transplantation 2001;72:377–384.

19   Larsen CP, Elwood ET, Alexander DZ, Ritchie SC, Tucker-Burden C, Cho HR, Aruffo A, Hollenbaugh D, Linsley PS, Winn KJ, Pearson TC: Long-term acceptance of skin and cardiac allografts after blocking CD40 and CD28 pathways. Nature 1996;381:434–438.

20   Vincenti F: What's in the pipeline? New immunosuppressive drugs in transplantation. Am J Transplant 2002;2:898–903.

21   Hausen B, Klupp J, Christians U, Higgins JP, Baumgartner RE, Hook LE, Friedrich S, Celnicker A, Morris RE: Coadministration of either cyclosporine or steroids with humanized monoclonal antibodies against CD80 and CD86 successfully prolong allograft survival after life supporting renal transplantation in cynomolgus monkeys. Transplantation 2001;72:1128–1137.

22   Vincenti F, Muehlbacher F, Nashan B, Larsen C, Atillasoy E, Natarajan K, Charpentier B, LEA29Y study group: Co-stimulation blockade with LEA29Y in a calcineurin inhibitor-free maintenance regimen: 6 month efficacy and safety. Am J Transplant 2004;in press.

23   Nashan B, Grinyo J, Vincenti F, Halloran P, Hagerty D, Zhou W, Charpentier B, LEA29Y study group: Co-stimulation blockade with LEA29Y in renal transplant: Improved renal function and CV/metabolic profile at 6 months compared with cyclosporine. Am J Transplant 2004;in press.

24   Adams AB, Shiragugi N, Durham MM, Strobert E, Anderson D, Rees P, Coan S, Xu H, Blinder Y, Cheung M, Hollenbaugh D, Kenyon NS, Pearson TC: Calcineurin inhibitor-free CD28 blockade-based protocol protects allogeneic islets in nonhuman primates. Diabetes 2002;51:1–6.

Flavio Vincenti, MD
University of California, San Francisco, Kidney Transplant Service
505 Parnassus Avenue, Rm 884M
San Francisco, California 94143–0780
Tel. +1 415 476 4496, Fax +1 415 353 8974, E-Mail vincentif@surgery.ucsf.edu

Ronco C, Chiaramonte S, Remuzzi G (eds): Kidney Transplantation: Strategies to Prevent
Organ Rejection. Contrib Nephrol. Basel, Karger, 2005, vol 146, pp 30–42

Steroid or Calcineurin Inhibitor-Sparing Immunosuppressive Protocols

Josep M. Grinyó, Josep M. Cruzado

Servei de Nefrologia, Hospital Universitari de Bellvitge, University of Barcelona,
C. Feixa Llarga s/n, L'Hospitalet, Barcelona, Spain

Abstract

Steroids have accompanied other immunosuppressants throughout the history of renal transplantation. However, its permanent use has been associated with a myriad of adverse effects, which especially increase the already high cardiovascular risk of renal transplant patients. Nevertheless, steroid-sparing strategies may increase the risk of acute and chronic rejection that may worsen the fate of transplant recipients. The advent of new immunosuppressants have renovated the interest on steroid-sparing protocols, and the results of the new trials suggest that these strategies may be safe enough in view of the low rates of acute rejection and stable renal function reported.

On the other hand, calcineurin inhibitors (CNIs) have been considered the cornerstone of transplant immunosuppression though their nephrotoxicity has been one of the major clinical problems in the use of these immunosuppressants. The balance between preventing immunological allograft losses and the management of CNI-related nephrotoxicity is still an issue in renal transplantation. CNI reduction or elimination may increase the risk of acute and chronic rejection. Because of these concerns, in most instances CNI have been used at conventional doses in induction and maintenance therapy. As in the case of steroid-sparing strategies, the new therapeutic arsenal has provided a new impulse in CNI-sparing regimens, with an acceptable low rate of acute rejection, well-preserved renal function and without an apparent increased risk of chronic rejection, which may pave the way for a new era in immunosuppression.

In the last years the aim in the majority of immunosuppressive regimens was to reduce the incidence and severity of acute rejection, because it was considered a deleterious prognostic factor for graft outcome. In the last decade, triple therapy consisting of a CNI, anti-metabolite, and steroids was widely used in many renal transplant centers, as induction and maintenance regimens. In this period, nearly all kidney transplant recipients received corticosteroid therapy prior to discharge, although the proportion of patients receiving steroids declined slightly at the end this period [1]. This tendency may reflect the concern in the transplant community of the importance of steroid-related morbidity in transplant patients. On the other hand, the permanent use of the nephrotoxic CNIs may induce chronic renal failure, even in transplant recipients of nonrenal solid allografts [2], apart from other well-known adverse effects of these agents such as hypertension and hyperlipidemia, which may increase the cardiovascular risk of the transplant population. As a consequence different attempts have been undertaken to spare steroids and CNIs to reduce comorbidity in kidney transplant patients.

Steroid-Sparing Protocols

The main questions that arise in steroid-sparing protocols are patient's selection, the timing after transplantation and the concomitant immunosuppression.

Initial reports, in the so-called cyclosporine (CsA) era, alerted on the increased risk of acute rejection and after steroid withdrawal in renal transplantation in patients treated with calcineurin-inhibitor (CNI) and azathioprine (AZA) [3]. In pediatric patients receiving CsA, stopping steroids was followed by 56% rate of acute rejection episodes [4]. In a single center experience with 100 patients, early discontinuation, black race, and renal function were identified as risk factors for subsequent rejection episodes after steroid withdrawal [5]. In a multicenter randomized and double-blind, placebo-controlled Canadian trial in 523 patients under CsA therapy, prednisone discontinuation at 90 days after transplantation significantly reduced actuarial 5-year graft survival rates to 73% in comparison to 85% in patients who remained with prednisone [6]. Results of a meta-analysis suggested that avoiding steroid therapy from the time of transplantation or withdrawing steroid therapy at some time after transplantation increased the risk of acute allograft rejection without adversely affecting patient or graft survival [7]. As expected, the beneficial effects of maintenance immunosuppression off steroids were catch-up growth in children [8] and reduced incidence of hypertension, improved glycemic control, and reduced total levels of serum lipids, although the long-term consequences of steroid-free CsA-based immunosuppression in renal transplantation were not clear in these studies reported in the mid nineties [9]. In a prospective and randomized trial in 100 established patients 1–6 years after transplantation treated with CsA and AZA, steroid discontinuation, over about 4 months, resulted in a rise in mean plasma creatinine at the end of the withdrawal period and at 2 and 3 years from trial entry. Changes in several clinical and metabolic indices were also observed in association with steroid withdrawal. Blood pressure declined but the reduction was incompletely sustained, being more evident immediately after steroid withdrawal than at one year. The data from this trial indicated that steroid withdrawal was feasible in most patients with stable graft function on triple immunosuppression and had potential beneficial metabolic effects although a substantial proportion of patients showed a reduction in graft function, indicating a need for caution in considering the long term outcome [10]. In another prospective trial comparing CsA monotherapy with AZA-prednisone maintenance therapy from 3 months after transplantation, the incidence of rejection within 3 months after the start of steroid withdrawal or conversion from CsA to AZA was 30 and 25% respectively, but at 2 years after transplantation, serum creatinine levels were significantly lower in the AZA-prednisone group than in the CsA group, although there were no differences in graft survival at 5 years after transplantation [11]. Late steroid withdrawal in patients treated with CsA with stable renal function at least one year after kidney transplantation showed that acute rejection was the main cause of withdrawal failure (26%), although no grafts were lost due to rejection [12]. Beneficial effects were found regarding hypertension, hypercholesterolemia, hyperglycemia, and appearance. In an Italian prospective study [13], the efficacy of AZA adjuncts to CsA at the time of steroid withdrawal, 6 months after transplantation, versus CsA monotherapy, in preventing acute rejection was compared. Steroid resumption because of acute rejection was significantly higher in the CsA monotherapy group than in the CsA-AZA group (57 vs. 29%), although serum creatinine did not differ, and graft survival was similar in both groups. Other studies on gradual withdrawal of steroids in the course of 6 months in patients receiving therapy with CsA and AZA, was associated with a low rate of acute rejection [14]. These results suggest that the addition of an anti-metabolite to a CNI may allow patients to remain steroid free. In a meta-analysis on randomized, controlled trials of immunosuppression withdrawal that comprised studies reported during the nineties with more than 1,400 patients, prednisone withdrawal entailed an increased risk of rejection by 14% and of late graft failure by 40% [15]. In contrast, CsA withdrawal in selected patients seemed to impart little risk of long-term graft failure. The conclusions of this meta-analysis may raise concerns about the safety of steroid-sparing strategies. In accordance, the published European Best Practice Guidelines for Renal Transplantation on late steroid or CsA withdrawal emphasized that steroid withdrawal is safe only in a proportion of graft recipients and is recommended only in low-risk patients, and the efficacy of the remaining immunosuppression should be considered (B). Moreover, the Guidelines also recommend that after steroid withdrawal, graft function has to be monitored very carefully because of the risk of a delayed but continuous loss of function due to chronic graft dysfunction, and that in the case of functional deterioration or dysfunction, steroids should be readministered (C) [16].

Nevertheless, the introduction of new xenobiotic immunosuppressants, such as mycophenolate mofetil (MMF) and biological agents, able to reduce to very low rates the incidence of acute rejection in conjunction with CNIs, has renovated the interest for steroid-sparing strategies in the recent years. In an open pilot study in our center with a small number of low-risk patients treated with CsA and steroids, prednisone elimination was not accompanied by rejection episodes, and renal function remained stable, suggesting that corticosteroids could be safely and successfully withdrawn from renal allograft recipients receiving CsA and MMF [17]. The feasibility of steroid-sparing strategies in patients treated with MMF has been intensively explored in the last few years. Following the criteria that a later steroid withdrawal was safer than an early elimination, steroid withdrawal under MMF therapy in prospective and controlled trials was initially attempted beyond 3 months after transplantation. More recently steroids have been stopped the first few days after transplantation or even completely avoided.

In an European multicenter, randomized, double-blind, 6-month, controlled steroid dose-reduction study in 500 renal transplant recipients with an unblind 6-month follow-up with a low/stop arm, corticosteroids were given at half the dosage of control for 3 months from the date of transplantation, and then withdrawn [18]. The comparator group received conventional doses of steroids. Both arms were given CsA and MMF. At 6 months the low/stop group had significantly more biopsy-proven acute rejection episodes than the control (23 vs. 14%) and at 12 months this increased to 25 vs. 15%. However, most rejections were Banff grade I, and renal function remained similar in both groups, and graft loss at month 12 was 5% in the low/stop group versus 4% in the control. The lipid profile, bone mineral density and blood pressure were better in patients off steroids. This first large study with MMF indicated that reduction and withdrawal of the prophylactic corticosteroid dose was feasible without an unacceptable increase in serious rejection episodes. In contrast, in a similar trial conducted in the USA, with a similar sample size, the recruitment was stopped after half of the patients were enrolled, because of excess rejection in the steroid-withdrawal group [19]. A careful review of this study showed that the high incidence of acute rejection was mainly restricted to African American patients, and that Caucasians had a similar rate of rejection to that observed in the previous European study. In a recently published uncontrolled study in which African American transplant patients were initially treated with sirolimus, tacrolimus, and corticosteroids, prednisone was withdrawn from eligible patients free of acute rejection beginning as early as 3 months post-transplant [20]. 7% of these patients developed acute rejection and at last follow-up, 27 of 30 patients

(90%) remained steroid free. However, there was a statistically insignificant trend toward an increased serum creatinine concentration before and after steroid withdrawal, which maintains the concerns on steroid withdrawal in African American patients despite the potent immunosuppressive regimen used in this trial consisting of these two macrolides.

The use of MMF may give the opportunity to compare CNI withdrawal and steroids cessation in the same trial. This innovative approach has been explored in a Dutch study [21]. At 6 months after transplantation, 212 patients were randomized to stop CsA, stop prednisone, or continue triple drug therapy with these two agents and MMF. Interestingly, patients off steroids experienced a similar incidence of acute and chronic rejection than the triple therapy group and less than those without CsA 2 years after transplant surgery.

Biological immunosuppressants, anti-IL-2-R monoclonal antibodies or polyclonal preparations may also help to design steroid-sparing strategies. Aiming to minimize the toxicity of steroids and CNIs at the same time, in a single center study, daclizumab, low tacrolimus exposure, MMF and steroid discontinuation was compared to a conventional regimen with tacrolimus, MMF and steroids [22]. In this trial, patients free of steroids experienced significantly less acute rejections, and graft function at one year was significantly better than those in the triple therapy group. In 51 living related kidney transplantation, polyclonal rabbit anti-thymocyte globulin has been employed in conjunction with MMF and CsA for rapid discontinuation of steroids 5 days after transplantation. With this immunosuppressive protocol, 87% of the patients evolved free of acute rejection at 12 months after transplantation, with an acceptable renal function and no differences with respect to historical controls treated with triple therapy without induction [23]. In a similar approach, basiliximab added to a maintenance regimen consisting of CsA and MMF mofetil was studied for its effectiveness in allowing early corticosteroid withdrawal at 4 days after transplantation in de novo renal allograft recipients. The incidence of biopsy-proven acute rejection at 12 months was not significantly different between the steroid-withdrawal group (20%) and the standard treatment group (16%), and renal function remained stable and similar in both groups at the end of the first year [24]. In the same direction, a prospective multicenter study investigated whether it is feasible to withdraw steroids early after transplantation with the use of anti-IL-2-R α induction, tacrolimus and MMF [25]. A total of 364 patients were randomized to receive either two doses of daclizumab and, for the first 3 days, high doses of prednisolone or steroids (tapered to 0 mg at week 16). All patients received tacrolimus and MMF. The incidence of biopsy-confirmed acute rejection at 12 months was not different between the daclizumab group (15%) and the controls (14%) and graft survival at 12 months was comparable in the two groups. These last studies suggest that regimens without steroids are close in renal transplantation. This therapeutic alternative has been investigated in pediatric population that could benefit most of steroid-free protocols. In a complete steroid-free immunosuppressive regimen in 100 children with an initial 10-day anti-thymocyte induction, and maintenance therapy with CsA and MMF, the rate of acute rejection was low (13%), and 4-year graft survival of 82% [26]. In a small single center study with 10 low-risk pediatric transplant recipients, steroids were substituted with extended daclizumab use, in combination with tacrolimus and MMF and compared to steroid-based historical controls [27]. In this preliminary report, there were no clinical acute rejection episodes and protocol biopsies did not display signs of chronic rejection. Besides the benefits on metabolic profile and cosmetic appearance, patients in the steroid-free regime did not require anti-hypertensive drugs and optimized renal function and growth.

Steroid discontinuation may also affect the pharmacokinetics of MMF. Steroid withdrawal is followed by an increase of mycophenolic acid exposure due to a decrease in its clearance. This is attributed to the reversion of the enhanced activity of uridine diphosphate-GT, responsible for mycophenolic acid metabolism, induced by steroid [28]. This increased exposure to the acid could compensate the interruption of steroid treatment.

CNI-Sparing Protocols

The balance between preventing immunological allograft losses and the management of CNI-related nephrotoxicity is still an issue in renal transplantation. CNI reduction or elimination may increase the risk of acute and chronic rejection. Because of these concerns, in most instances, CNI have been used at conventional doses in induction and maintenance therapy. However, historical reports from the first times of the CsA era showed that conversion from CsA to AZA at one year after renal transplantation resulted in improvement in both blood pressure control and renal allograft function, and was not associated with significant adverse effects on long-term patient or graft survival, despite an increased incidence of acute rejection within the first few months after conversion [29].

With the introduction of potent xenobiotic and biological immunosuppressants, three main CNI-sparing strategies have been investigated: CNI minimization, CNI withdrawal and complete avoidance of CNI.

CNI Minimization
The advent of MMF enhanced CsA doses reduction to ameliorate renal function in established patients with chronic renal allograft dysfunction. In patients with progressive deterioration of renal function, the addition of MMF

and the reduction of CsA doses by 50% resulted in a short-term improvement in renal function [30]. In our center, we did a similar therapeutic maneuver targeting CsA levels between 40–60 ng/ml, which was followed by improvement in renal function, reduction of TGF-β1 production, and improved the control of hypertension, without increasing the incidence of acute rejection [31], in the short term. In another trial, in 118 renal transplant recipients with declining kidney function and biopsy-proven chronic allograft nephropathy, CNIs (CsA or tacrolimus) dose was reduced or discontinued with either the addition or continuation of MMF and low-dose steroids. The long-term evaluation of these treatment changes showed that 72.2% of the CNI-withdrawal patients, 54.4% of reduced-dose CsA group, and 40% of the reduced-dose tacrolimus group had improved the slope of decay of renal function or lack of deterioration, suggesting that the reduction and possible withdrawal of CNIs may be necessary to slow the rate of loss of renal function in patients with chronic allograft nephropathy and deteriorating renal function [32].

CNI Withdrawal

CNI withdrawal has been attempted in regimens containing MMF or sirolimus (SRL). In a large prospective and randomized trial, in stable patients treated with MMF, CsA and steroids, CsA discontinuation was accompanied by a modest amelioration of renal function in the short term, and a better lipid profile, although the incidence of acute rejection after CsA withdrawal was 11% in comparison to 2.4% in patients who remained under CsA, but without graft losses [33]. The effect of adding MMF to the maintenance immunosuppression followed by CsA withdrawal was studied in 143 unstable recipients with deteriorating renal function and biopsy proven chronic allograft dysfunction in a multicenter randomized controlled trial [34]. In patients treated with MMF and CsA withdrawal, renal function stabilized or significantly improved in 58% of them, compared to 28% in the CsA-treated patients. In contrast with the previous study, only 1.5% of MMF patients experienced acute rejection, and 3% in the CsA-treated patients at 34 weeks after the initiation of the study. These preliminary results indicate that in patients with creeping creatinine, the substitution of CsA with MMF may be a safe therapeutic option. In a smaller single center trial recently reported, the potential benefits on graft function by the introduction of MMF with or without CNI withdrawal has been studied in long-term transplant recipients with histologically proven chronic allograft nephropathy and deteriorating renal function [35]. Renal function significantly improved in 19 patients with MMF/CNI withdrawal compared with 20 with MMF/CNI continuation, with a better control of blood pressure in the former group, and no rejections occurred during the 34-month follow-up period.

SRL, an m-TOR inhibitor, is a potent immunosuppressant that inhibits cell proliferation driven by growth factor. When associated with CNI, SRL greatly reduces the incidence of acute rejection [36], and associated with anti-metabolites, either AZA or MMF, prevents acute rejection to similar rates than CNI [37, 38]. Considering the potency of SRL, two multicenter trials were conducted to evaluate the efficacy and safety of CsA withdrawal 3 months after transplantation in comparison with a maintenance immunosuppression based on CsA, SRL and steroids [39, 40]. CsA withdrawn was accompanied by a concentration-controlled increase in SRL doses. In the first study with 197 patients, in the intention-to-treat population, rates of biopsy-confirmed acute rejection at 12 months were not significantly different between groups with or without CsA (18.6 vs. 22.0%, respectively). In the second larger study, in 430 eligible patients treated with SRL, CsA discontinuation was followed by a sustained improvement in renal function at 3 years, although numerically more biopsy-proven acute rejections occurred after CsA elimination in comparison with the continued use of CsA and SRL (5.6 vs. 10.2%) but without a negative impact on graft survival [41]. Moreover, in protocol-mandated biopsies at engraftment and at 12 and 36 months, it was observed that the histological score at 36 months among patients with serial biopsies was significantly lower with SRL-steroids than with SRL-CsA-steroids, and inflammation and the tubular atrophy scores decreased significantly in the SRL-steroid group between 12 and 36 months [42]. These histological findings reveal that the elimination of CsA in SRL-treated recipients does not increase the risk of chronic rejection. In a multicenter trial with a similar design than these CsA withdrawal studies, this maneuver was compared with CsA dose minimization in conjunction with SRL. As in the previous studies the better renal function was obtained in the CsA-withdrawal patients [43]. Tacrolimus elimination has also been assayed in SRL-based regimens. We recently performed a randomized trial to compare two regimens in 87 low-risk kidney allograft recipients in the first year after transplantation [44]. Both regimens initially included sirolimus, tacrolimus and steroids; one with long-term maintenance with these drugs versus tacrolimus withdrawal. Both macrolide doses were adjusted to reach target levels. In one group, SRL was used at reduced doses and levels and tacrolimus at conventional doses, and vice versa in the other group with tacrolimus elimination beyond 3 months after transplantation. Both groups displayed a low rate of acute rejection and the analysis of patients on therapy showed better renal function, and lower blood pressure in patients who withdrew tacrolimus.

CNI Avoidance

Because suboptimal organs may be more susceptible to CNI nephrotoxicity, pilot studies free of these agents with polyclonal antibodies and MMF were conducted in the second half of the nineties. Twelve patients older than 50 years, and receiving a renal graft from a donor older than 50 years, were treated primarily with MMF combined with steroids and an induction therapy using anti-thymocyte globulin, and without the addition of CsA [45]. With this regimen the incidence of acute rejection was low (8.3%), renal function was preserved at 6 months, although the rate of cytomegalovirus infections was high (41%). In an attempt to avoid the use of cyclosporine, we carried out a prospective study in low-immunological risk recipients of suboptimal kidneys, using an immunosuppressive protocol combining rabbit anti-thymocyte globulin in induction with a bi-therapy of MMF and steroids [46]. Half of the patients received CNI, 7 (24%) out of 30 patients developed acute rejection, and renal function was acceptable at 5 years, but the rate of opportunistic infection and neoplasia was high with this regimen.

A refinement of the CNI-free polyclonal-based regimens would be the use of anti-IL-2-R monoclonal antibodies. However, avoidance of CNI in a single-arm study with MMF daclizumab and steroids resulted in an unacceptable high incidence of acute rejection, although with a good renal function and patient and graft survival [47]. The addition of SRL to this immunosuppressant combination has greatly reduced the incidence of acute rejection. In a single center experience [48], the association of basiliximab, MMF and CsA or SRL, efficiently prevented acute rejection, but the SRL-treated patients had a significantly higher creatinine clearance than patients given CsA and in 2-year protocol biopsies a much lower rate of chronic allograft nephropathy (37 vs. 78%).

CNI-free regimens may reinvigorate in tolerogenic protocols, because in contrast with SRL, the treatment with CsA abrogates the tolerogenic effect of costimulation blockade by inhibiting T-cell proliferation and apoptosis [49]. The use of Campath, a chimeric anti-CD52 monoclonal antibody, that depletes lymphocytes B cells and monocytes, in combination with rapamycin maintenance monotherapy, have been studied in a pilot study with 29 patients. Twenty seven per cent developed acute rejection, and a high proportion had pathological evidence of a humoral component of their rejection.

Finally, costimulation blockade with second generation CTLA4Ig fusion proteins (LEA29Y), with a high avidity for CD86 and CD80, may help in the development of safe CNI-free regimens. In a multicenter trial, a therapy with CsA, MMF, steroids plus basiliximab was compared to LEA29Y in conjunction with MMF, steroids and basiliximab. Both treatment arms had a similar low incidence of acute rejection (18 and 19%) at 6 months, and with a more favorable cardiovascular risk profile in the costimulation blockade arm without CsA [50, 51].

In summary, the new therapeutic arsenal in renal transplantation has allowed to enter in a new era of low comorbidity protocols aimed to avoid drug-related adverse effects and improve graft and patient survivals. The next challenge would be to conciliate in the same protocol steroid and CIN-sparing strategies. The real impact of steroid-sparing or CNI-sparing regimens on graft and patient outcome will need a close follow-up in the long term.

References

1   Helderman JH, Bennett WM, Cibrik DM, Kaufman DB, Klein A, Takemoto SK: Immunosuppression: Practice and trends. Am J Transplant 2003;3(suppl 4):41–52.
2   Ojo AO, Held PJ, Port FK, Wolfe RA, Leichtman AB, Young EW, Arndorfer J, Christensen L, Merion RM: Chronic renal failure after transplantation of a nonrenal organ. N Engl J Med 2003; 349:931–940.
3   Stratta RJ, Armbrust MJ, Oh CS, Pirsch JD, Kalayoglu M, Sollinger HW, Belzer FO: Withdrawal of steroid immunosuppression in renal transplant recipients. Transplantation 1988;45:323–328.
4   Reisman L, Lieberman KV, Burrows L, Schanzer H: Follow-up of cyclosporine-treated pediatric renal allograft recipients after cessation of prednisone. Transplantation 1990;49:76–80.
5   Hricik DE, Whalen CC, Lautman J, Bartucci MR, Moir EJ, Mayes JT, Schulak JA: Withdrawal of steroids after renal transplantation – Clinical predictors of outcome. Transplantation 1992;53:41–45.
6   Sinclair NR: Low-dose steroid therapy in cyclosporine-treated renal transplant recipients with well-functioning grafts. The Canadian Multicentre Transplant Study Group. CMAJ 1992;147: 645–657.
7   Hricik DE, O'Toole MA, Schulak JA, Herson J: Steroid-free immunosuppression in cyclosporine-treated renal transplant recipients: A meta-analysis. J Am Soc Nephrol 1993;4:1300–1305.
8   Chao SM, Jones CL, Powell HR, Johnstone L, Francis DM, Becker GJ, Walker RG: Triple immunosuppression with subsequent prednisolone withdrawal: 6 years' experience in paediatric renal allograft recipients. Pediatr Nephrol 1994;8:62–69.
9   Schulak JA, Hricik DE: Steroid withdrawal after transplantation. Clin Transplant 1994;8:211–216.
10   Ratcliffe PJ, Dudley CR, Higgins RM, Firth JD, Smith B, Morris PJ: Randomised controlled trial of steroid withdrawal in renal transplant recipients receiving triple immunosuppression. Lancet 1996;348:643–648.
11   Hilbrands LB, Hoitsma AJ, Koene KA: Randomized, prospective trial of cyclosporine monotherapy versus azathioprine-prednisone from three months after renal transplantation. Transplantation 1996;61:1038–1046.
12   Hollander AA, Hene RJ, Hermans J, van Es LA, van der Woude FJ: Late prednisone withdrawal in cyclosporine-treated kidney transplant patients: A randomized study. J Am Soc Nephrol 1997;8: 294–301.
13   Sandrini S, Maiorca R, Scolari F, Cancarini G, Setti G, Gaggia P, Cristinelli L, Zubani R, Bonardelli S, Maffeis R, Portolani N, Nodari F, Giulini SM: A prospective randomized trial on azathioprine addition to cyclosporine versus cyclosporine monotherapy at steroid withdrawal, 6 months after renal transplantation. Transplantation 2000;69:1861–1867.
14   Matl I, Lacha J, Lodererova A, Simova M, Teplan V, Lanska V, Vitko S: Withdrawal of steroids from triple-drug therapy in kidney transplant patients. Nephrol Dial Transplant 2000;15:1041–1045.
15   Kasiske BL, Chakkera HA, Louis TA, Ma JZ: A meta-analysis of immunosuppression withdrawal trials in renal transplantation. J Am Soc Nephrol 2000;11:1910–1917.
16   EBPG Expert Group on Renal Transplantation. European best practice guidelines for renal transplantation. IV. Long-term management of the transplant recipient. IV.3.1 Long-term immunosuppression. Late steroid or cyclosporine withdrawal. Nephrol Dial Transplant 2002; 17(suppl 4):19–20.
17   Grinyo JM, Gil-Vernet S, Seron D, Cruzado JM, Moreso F, Fulladosa X, Castelao AM, Torras J, Hooftman L, Alsina J: Steroid withdrawal in mycophenolate mofetil-treated renal allograft recipients. Transplantation 1997;63:1688–1690.

18 Vanrenterghem Y, Lebranchu Y, Hene R, Oppenheimer F, Ekberg H: Double-blind comparison of two corticosteroid regimens plus mycophenolate mofetil and cyclosporine for prevention of acute renal allograft rejection. Transplantation 2000;70:1352–1359.

19 Ahsan N, Hricik D, Matas A, Rose S, Tomlanovich S, Wilkinson A, Ewell M, McIntosh M, Stablein D, Hodge E: Prednisone withdrawal in kidney transplant recipients on cyclosporine and mycophenolate mofetil – A prospective randomized study. Steroid Withdrawal Study Group. Transplantation 1999;68:1865–1874.

20 Hricik DE, Knauss TC, Bodziak KA, Weigel K, Rodriguez V, Seaman D, Siegel C, Valente J, Schulak JA: Withdrawal of steroid therapy in African American kidney transplant recipients receiving sirolimus and tacrolimus. Transplantation 2003;76:938–942.

21 Smak Gregoor PJ, de Sevaux RG, Ligtenberg G, Hoitsma AJ, Hene RJ, Weimar W, Hilbrands LB, van Gelder T: Withdrawal of cyclosporine or prednisone six months after kidney transplantation in patients on triple drug therapy: A randomized, prospective, multicenter study. J Am Soc Nephrol 2002;13:1365–1373.

22 Kuypers DR, Evenepoel P, Maes B, Coosemans W, Pirenne J, Vanrenterghem Y: The use of an anti-CD25 monoclonal antibody and mycophenolate mofetil enables the use of a low-dose tacrolimus and early withdrawal of steroids in renal transplant recipients. Clin Transplant 2003; 17:234–241.

23 Matas AJ, Ramcharan T, Paraskevas S, Gillingham KJ, Dunn DL, Gruessner RW, Humar A, Kandaswamy R, Najarian JS, Payne WD, Sutherland DE: Rapid discontinuation of steroids in living donor kidney transplantation: A pilot study. Am J Transplant 2001;1:278–283.

24 Vincenti F, Monaco A, Grinyo J, Kinkhabwala M, Roza A: Multicenter randomized prospective trial of steroid withdrawal in renal transplant recipients receiving basiliximab, cyclosporine microemulsion and mycophenolate mofetil. Am J Transplant 2003;3:306–311.

25 ter Meulen CG, van Riemsdijk I, Hene RJ, Christiaans MH, Borm GF, van Gelder T, Hilbrands LB, Weimar W, Hoitsma AJ: Steroid-withdrawal at 3 days after renal transplantation with anti-IL-2 receptor alpha therapy: A prospective, randomized, multicenter study. Am J Transplant 2004;4: 803–810.

26 Birkeland SA: Steroid-free immunosuppression in renal transplantation: A long-term follow-up of 100 consecutive patients. Transplantation 2001;71:1089–1090.

27 Sarwal MM, Yorgin PD, Alexander S, Millan MT, Belson A, Belanger N, Granucci L, Major C, Costaglio C, Sanchez J, Orlandi P, Salvatierra O Jr: Promising early outcomes with a novel, complete steroid avoidance immunosuppression protocol in pediatric renal transplantation. Transplantation 2001;72:13–21.

28 Cattaneo D, Perico N, Gaspari F, Gotti E, Remuzzi G: Glucocorticoids interfere with mycopheno-late mofetil bioavailability in kidney transplantation. Kidney Int 2002;62:1060–1067.

29 MacPhee IA, Bradley JA, Briggs JD, Junor BJ, MacPherson SG, McMillan MA, Rodger RS, Watson MA: Long-term outcome of a prospective randomized trial of conversion from cyclosporine to azathioprine treatment one year after renal transplantation. Transplantation 1998;66:1186–1192.

30 Weir MR, Anderson L, Fink JC, Gabregiorgish K, Schweitzer EJ, Hoehn-Saric E, Klassen DK, Cangro CB, Johnson LB, Kuo PC, Lim JY, Bartlett ST: A novel approach to the treatment of chronic allograft nephropathy. Transplantation 1997;64:1706–1710.

31 Hueso M, Bover J, Seron D, Gil-Vernet S, Sabate I, Fulladosa X, Ramos R, Coll O, Alsina J, Grinyo JM: Low-dose cyclosporine and mycophenolate mofetil in renal allograft recipients with suboptimal renal function. Transplantation 1998;66:1727–1731.

32 Weir MR, Ward MT, Blahut SA, Klassen DK, Cangro CB, Bartlett ST, Fink JC: Long-term impact of discontinued or reduced calcineurin inhibitor in patients with chronic allograft nephropathy. Kidney Int 2001;59:1567–1573.

33 Abramowicz D, Manas D, Lao M, Vanrenterghem Y, Del Castillo D, Wijngaard P, Fung S: Cyclosporine withdrawal from a mycophenolate mofetil-containing immunosuppressive regimen in stable kidney transplant recipients: A randomized, controlled study. Cyclosporine Withdrawal Study Group. Transplantation 2002;74:1725–1734.

34 Dudley CRK: The MMF 'Creeping Creatinine' Study Group. MMF substitution for CsA is an effective and safe treatment of chronic allograft dysfunction: Results of a multi-center randomized controlled trial. Washington, American Transplant Congress, May 2002.

35   Suwelack B, Gerhardt U, Hohage H: Withdrawal of cyclosporine or tacrolimus after addition of mycophenolate mofetil in patients with chronic allograft nephropathy. Am J Transplant 2004;4: 655–662.

36   Kahan BD: Efficacy of sirolimus compared with azathioprine for reduction of acute renal allograft rejection: A randomised multicentre study. The Rapamune US Study Group. Lancet 2000;356: 194–202.

37   Groth CG, Backman L, Morales JM, Calne R, Kreis H, Lang P, Touraine JL, Claesson K, Campistol JM, Durand D, Wramner L, Brattstrom C, Charpentier B: Sirolimus (rapamycin)-based therapy in human renal transplantation: Similar efficacy and different toxicity compared with cyclosporine. Sirolimus European Renal Transplant Study Group. Transplantation 1999;67: 1036–1042.

38   Kreis H, Cisterne JM, Land W, Wramner L, Squifflet JP, Abramowicz D, Campistol JM, Morales JM, Grinyo JM, Mourad G, Berthoux FC, Brattstrom C, Lebranchu Y, Vialtel P: Sirolimus in association with mycophenolate mofetil induction for the prevention of acute graft rejection in renal allograft recipients. Transplantation 2000;69:1252–1260.

39   Gonwa TA, Hricik DE, Brinker K, Grinyo JM, Schena FP: Randomized trial of tacrolimus in combination with sirolimus or mycophenolate mofetil in kidney transplantation: Results at 6 months. Sirolimus Renal Function Study Group. Transplantation 2003;75:1213–1220.

40   Kreis H, Oberbauer R, Campistol JM, Mathew T, Daloze P, Schena FP, Burke JT, Brault Y, Gioud-Paquet M, Scarola JA, Neylan JF: Long-term benefits with sirolimus-based therapy after early cyclosporine withdrawal. Rapamune Maintenance Regimen Trial. J Am Soc Nephrol 2004;15: 809–817.

41   Oberbauer R, Kreis H, Johnson RW, Mota A, Claesson K, Ruiz JC, Wilczek H, Jamieson N, Henriques AC, Paczek L, Chapman J, Burke JT: Long-term improvement in renal function with sirolimus after early cyclosporine withdrawal in renal transplant recipients: 2-year results of the Rapamune Maintenance Regimen Study. Rapamune Maintenance Regimen Study Group. Transplantation 2003;7:364–370.

42   Mota A, Arias M, Taskinen EI, Paavonen T, Brault Y, Legendre C, Claesson K, Castagneto M, Campistol JM, Hutchison B, Burke JT, Yilmaz S, Hayry P, Neylan JF: Sirolimus-based therapy following early cyclosporine withdrawal provides significantly improved renal histology and function at 3 years. Rapamune Maintenance Regimen Trial. Am J Transplant 2004;4: 953–961.

43   Baboolal K: A phase III prospective, randomized study to evaluate concentration-controlled sirolimus (rapamune) with cyclosporine dose minimization or elimination at six months in de novo renal allograft recipients. Transplantation 2003;75:1404–1408.

44   Grinyó JM, JM, Campistol JM, Paul J, García-Martínez J, Morales JM, Prats D, Arias M, Brunet M, Cabrera J, Granados E: Pilot randomized study of early tacrolimus withdrawal from a regimen with sirolimus plus tacrolimus in kidney transplantation. Am J Transplant 2004;4:1–7.

45   Zanker B, Schneeberger H, Rothenpieler U, Hillebrand G, Illner WD, Theodorakis I, Stangl M, Land W: Mycophenolate mofetil-based, cyclosporine-free induction and maintenance immunosuppression: First-3-months analysis of efficacy and safety in two cohorts of renal allograft recipients. Transplantation 1998;66:44–49.

46   Grinyo JM, Gil-Vernet S, Cruzado JM, Caldes A, Riera L, Seron D, Rama I, Torras J: Calcineurin inhibitor-free immunosuppression based on antithymocyte globulin and mycophenolate mofetil in cadaveric kidney transplantation: Results after 5 years. Transpl Int 2003;16: 820–827.

47   Vincenti F, Ramos E, Brattstrom C, Cho S, Ekberg H, Grinyo J, Johnson R, Kuypers D, Stuart F, Khanna A, Navarro M, Nashan B: Multicenter trial exploring calcineurin inhibitors avoidance in renal transplantation. Transplantation 2001;7:1282–1287.

48   Flechner SM, Goldfarb D, Modlin C, Feng J, Krishnamurthi V, Mastroianni B, Savas K, Cook DJ, Novick AC: Kidney transplantation without calcineurin inhibitor drugs: A prospective, randomized trial of sirolimus versus cyclosporine. Transplantation 2002;74:1070–1076.

49   Li Y, Li XC, Zheng XX, Wells AD, Turka LA, Strom TB: Blocking both signal 1 and signal 2 of T-cell activation prevents apoptosis of alloreactive T cells and induction of peripheral allograft tolerance. Nat Med 1999;5:1298–1302.

50   Vincenti F, Muehlbacher F, Nashan B, Larsen C, Atillasoy E, Natarajan K, Charpentier B, LEA29Y study group: Co-stimulation blockade with LEA29Y in a calcineurin inhibitor-free maintenance regimen: 6 month efficacy and safety. Boston, American Transplant Congress, May 2004.
51   Nashan B, Grinyo JM, Vincenti F, Halloran P, Hagerty D, Zhou W, Charpentier B, LEA29Y study group: Co-stimulation blockade with LEA29Y in renal transplant: Improved renal function and CV/metabolic profile at 6 months compared with cyclosporine. Boston, American Transplant Congress, May 2004.

Josep M. Grinyó
Servei de Nefrologia, Hospital Universitari de Bellvitge
University of Barcelona, C. Feixa Llarga s/n, L'Hospitalet
ES–08907 Barcelona (Spain)
Tel. +34 932607604, Fax +34 932607607, E-Mail jgrinyo@csub.scs.es

Ronco C, Chiaramonte S, Remuzzi G (eds): Kidney Transplantation: Strategies to Prevent
Organ Rejection. Contrib Nephrol. Basel, Karger, 2005, vol 146, pp 43–53

....

Steroid-Free Lymphocyte Depletion Protocols

The Potential for Partial Tolerance?

Jerry McCauley

Thomas E. Starzl Transplantation Institute, University of Pittsburgh School of
Medicine, Pittsburgh, Pa., USA

Abstract

Induction of tolerance has been a longstanding goal in transplantation. Recent preliminary
studies using a steroid-free lymphocyte depletion strategy have been met with great excitement
and an equal degree of skepticism. Current studies in this area suggest that immunosuppression
can be reduced substantially but acute rejection develops at a rate approximately twice that of
standard triple drug protocols. None of the lymphocyte-depleting protocols has resulted in full
tolerance as evidenced by patients attaining a maintenance drug-free state. Workers in the field
have suggested that a degree of tolerance is achieved and they have coined a growing number
of terms to describe this state: prope tolerance, metastable tolerance, and partial tolerance. To
date, patient follow-up has been relatively short leaving many unanswered questions about graft
survival, chronic allograft nephropathy, and the minimally effective maintenance immunosuppression. Despite these limitations, steroid-free lymphocyte depletion may offer an exciting new
treatment paradigm.

Introduction

The advent of highly effective immunosuppressant agents over the past
two decades has made renal transplantation the treatment of choice for patients
with end-stage renal disease. Recent United Network of Organ Sharing reports
in the USA documented one-year graft survival rates of approximately 88% for
deceased donors and 95% for living donors [1]. Despite these encouraging
results, patients have continued to experience transplant-related infections and
other side effects of profound nonselective immunosuppression in addition to
the side effects of the drugs themselves. Animal models of renal transplantation

have demonstrated that long-term function of grafts without the need for immunosuppression (tolerance) is possible. Attempts at inducing tolerance in human subjects have frustrated the efforts of many workers. Brent [2] has described the quest for tolerance in humans as 'the search for the holy grail.' The topic of immunological tolerance has been reviewed recently by Brent [3, 4], Starzl et al. [5–7], Cosimi and Sachs [8] and others. Since this is a vast topic, this review will concentrate on those aspects directly germane to lymphocyte depletion as a strategy to induce tolerance.

Rationale for Use of Lymphocyte Depletion Protocols

Lymphocyte depletion as a method of inducing tolerance in renal transplantation is based upon extensive experimental studies in animals coupled with new strategies of immunosuppressant use after transplantation. Although lymphocyte depletion has played an important role in experimental tolerance induction over much of transplantation history, work by Knechtle et al. [9] was central in stimulating the recent flurry of clinical activity in this area. An anti-CD3 lymphocyte-depleting immunotoxin (FN18-CRM9) was administered to MHC-mismatched rhesus monkeys one week prior to renal transplantation. T lymphocyte depletion (by 2–3 logs) developed in peripheral and lymph nodes when measured 1–3 days later. These animals enjoyed graft survival greater than 100 days without any other immunosuppression. When donor skin grafts were placed 6 months later 5 of the 6 animals accepted the grafts but rejected third-party grafts. However, when one of the animals received a donor skin graft at 140 days (the earliest graft) it stimulated both rejections of the skin graft and the renal allograft. This experience demonstrated that lymphocyte depletion could induce tolerance but that it may be unstable and can be lost by an immunological stimulus administered early in the transplant course. This study stimulated human trials by Calne et al. [23–25] using Campath-1H described below.

In addition to the availability of agents capable of profound lymphocyte depletion, Starzl et al. [5–7] and Calne [13] have emphasized the importance of timing and intensity of immunosuppression as important factors in clinical tolerance induction. Since the initial studies of Cannon et al. [10] and Medawar and coworkers [11] in neonatal tolerance, a 'window of opportunity' for tolerance induction has been appreciated. For neonatal chicks receiving skin grafts, tolerance was induced only in animals transplanted within 3 days of birth. These pivotal studies generated the field of tolerance in transplantation. In adults receiving whole organ transplants, an early period in which tolerance may be easily induced may also exist.

Starzl et al. [5] has formulated a clinical approach to tolerance induction based upon lessons he learned during his vast experience in transplantation. His concept of mutual immunological engagement between donor and recipient was documented in a recent review [6]. Immunosuppression administered before transplantation may reduce the antigraft response by decreasing passenger leukocytes to a range in which clonal deletion may occur. Historical examples include the twin transplants by Murray and Hamburger using total body irradiation and other procedures such as thoracic duct drainage, and conventional antilymphocyte antibody preparations. In addition to pretreatment, avoidance of intense postsurgical immunosuppression is felt to be important in allowing donor-specific clonal deletion and tolerance. The early post-transplant period (the first, approximately 2 months) is theoretically vital in allowing tolerance to proceed [12]. These principles were the basis for his protocols of lymphocyte depletion described below.

Calne [13] has also developed a theoretical framework for clinical tolerance induction, which resulted in recent clinical efforts at tolerance induction with lymphocyte-depleting protocols. He argues that nonspecific immunosuppression impaired the mutual engagement between immunocompetent donor and recipient cells which was earlier described by Starzl. A 'window of opportunity for immunological engagement (WOFIE)' was proposed. From this approach, a short period (1–3 days) after surgery should be essentially free of immunosuppression to allow donor and recipient to engage. This approach has been tested in animal and human studies with encouraging results [14–17]. Calne suggests that '… elimination of aggressive T cell function should tip the balance in favor of a tolerant state' [13]. Lymphocyte depletion was proposed as the tool to tip the balance.

Both Starzl and Calne agree that minimization of immunosuppression after surgery may be important in clinical tolerance induction. Both also agree to the concept of a 'window of opportunity' although the timing appears to be somewhat different. These concepts in addition to robust depletion of lymphocytes form the basis for clinical studies described below.

Clinical Experience with Lymphocyte Depletion

Campath-1H

Campath-1H (Alemtuzumab), a CD52-specific monoclonal antibody, has become the most exciting agent in the growing clinical experience of steroid-free lymphocyte depletion protocols aimed at inducing some degree of tolerance. It profoundly depletes T-lymphocytes and reduces B cells and monocytes to a lesser degree [18]. This agent has been extensively studied in bone marrow transplantation and is indicated for the treatment of chronic lymphocytic leukemia [19–21]. Knechtle [22] has recently reviewed the clinical development of this agent.

Calne et al. [23, 24] generated the first clinical experience with Campath-1H in an attempt to induce tolerance. This experience was initially published as a research letter in Lancet in 1998 and subsequently his experience was expanded from the first 13 patients to 31 primary kidney transplant recipients. Two 20-mg doses of Campath-1H was administered to all patients. Five received the first dose before surgery and the remainder postoperatively. All patients received the second dose 24 hours after the initial dose. Twenty-four hours after surgery, all patients were given cyclosporine (Neoral) at a dose designed to reach targeted maintenance trough concentrations of 75–125 ng/ml. No maintenance steroids were planned but all patients were given 500 mg of methylprednisolone 30 min before the first Campath-1H dose to minimize cytokine release symptoms. Acute cellular rejections developed in 6 of 31 patients (19.4%) from 28 days to 13 months after surgery. Three patients required maintenance prednisolone and azathioprine due to rejection. Despite the high rejection rate most patients attained stable renal function with a steroid-free low dose cyclosporine monotherapy protocol. These results prompted Calne et al. to suggest that the patients had developed partial tolerance and he subsequently coined the term 'prope tolerance' which is a Latin term for almost. Watson et al. [25] provided the most recent update on this study at the American Transplant Congress in May 2004. The number of Campath-1H-treated patients had increased to 33 and a contemporaneous triple drug therapy group (cyclosporine, prednisone with Cellcept or sirolimus) was included as controls. One and 5-year graft survival was 94 and 79% for Campath-1H patients compared to 83 and 75% for controls (not significant). One patient in both groups had died from post-transplant lymphoproliferative disease. There was no difference in renal function or acute rejection but rejections developed later in the Campath-1H group compared to controls (170 vs. 16 days). No attempts were apparently made to wean or withdraw immunosuppression.

After the initial experience of Calne et al. two groups led by Knechtle from the University of Wisconsin and Kirk from the National Institute of Health in the USA designed studies to investigate different aspects of the potential of Campath-1H to induce tolerance. Kirk et al. [26] performed a study to specifically determine if depletion of peripheral and secondary T lymphocytes with Campath-1H was capable of inducing tolerance. Seven primary live-donor kidney transplant recipients were treated with Campath-1H without additional maintenance immunosuppression. Campath-1H was given intravenously at a dose of 0.3 mg/kg over a 3-hour period. Four patients received treatments on days −5, −3, and −1 and 2 patients were treated on days −3, −12, and +2.

One patient received pretreatment on days -1, $+1$, $+3$, and $+5$. Methylprednisolone (500 mg before dose 1, 125–250 mg before dose 2, and 60–125 mg before dose 3) was given to decrease cytokine release but all patients developed mild self-limited symptoms. Steroids were not planned thereafter. Follow-up at the time of this report was between 1.2 and 2.6 years.

This group has extensively studied the details of lymphocyte depletion in various cellular subsets. Peripheral lymphocyte depletion began within one hour of drug administration and repopulation of lymphocytes began at one month. Absolute lymphocyte counts failed to reach pretreatment levels at 18 months although at 6 months they had returned to normal values. Monocyte depletion was less severe and repopulation began within 3 weeks. Secondary lymphoid tissue clearance developed over 3–5 days and iliac lymph nodes revealed sparse medullary T cells. Previously activated (memory) cell phenotypes (CD3+, HLA DR+, CD45RO+) were less depleted than naïve cells. Nodal monocytes and B cells were unaffected. Natural killer cells were reduced but remained significantly present. Rejection and reperfusion prompted acute increases in monocytes. Neutrophils and platelet counts were not changed by therapy.

All the patients experienced acute rejection; 6 had clinical rejections 3–4 weeks after surgery and one had a subclinical rejection during this period. There was no change in absolute lymphocyte count preceding or during rejection. Five patients required bolus methylprednisolone followed by sirolimus therapy, one required OKT3 followed by sirolimus and one was treated with sirolimus alone. None of the patients remained off maintenance immunosuppression in the long term. The histology of rejection was characterized by primarily monocytic infiltration with a sparse contribution by the typical lymphocyte population. This study, in a very low immunological risk group, revealed that full tolerance was not achievable with Campath-1H pretreatment alone.

The Knechtle et al. [27, 28] approach was to use Campath-1H without steroids or calcineurin inhibitors but to use only low-dose sirolimus as maintenance therapy. This approach conceded that full tolerance was unlikely but attempted to minimize long-term use of immunosuppression. In this study, 29 primary renal transplant recipients were treated with Campath-1H and sirolimus without maintenance steroids. The first 24 patients received 20 mg of Campath-1H intravenously on the day of transplant (day 0) and a second 20 mg dose on day 1. All patients were treated with methylprednisolone 500 mg intravenously 30 min prior to Campath-1H. Sirolimus (2 mg orally) was given on the day after surgery and levels subsequently adjusted to reach targeted levels of 8–12 ng/ml. The last 5 patients were treated with Campath-1H on day -1 and were given one dose of Thymoglobulin (1.5 mg/kg i.v.) on day 1. A tapering dose of steroids was added to the latter patients with total daily doses beginning at 1,000 mg beginning on day -1 and discontinued by day 15. The sirolimus dosing was unchanged.

This study illustrated the limitations of lymphocyte depletion with Campath-1H when sirolimus is coadministered. The patients again experienced relatively high rates of acute rejection. Eight of 29 (28%) experienced acute cellular rejections and 5 of these had C4d+ humoral components. One graft was lost to rejection and the others with humoral components required combinations of plasmapheresis, Thymoglobulin, prednisone and rituximab to obtain control. Three of the 4 surviving grafts with humoral rejection were left with higher creatinine values compared to prerejection values. These investigators observed that recovering T cells were primarily CD52 negative (in normal subjects only 1–3% are CD52 negative). After the latter observation and the unacceptably high rejection rate, Thymoglobulin was added to the last 5 patients in an attempt to target the CD52 negative cells. Two of the 5 (40%) patients developed rejection; the first was Banff grade 1B (C4d negative) which required bolus steroid therapy, Thymoglobulin, plasmapheresis and IVIG to control. The second developed Banff IIA cellular rejection with thrombotic microangiopathy and was C4d positive. This patient required bolus steroids, Thymoglobulin and plasmapheresis and was later converted to tacrolimus, Cellcept and prednisone maintenance therapy. This group has recently reported preliminary results of an altered protocol using Campath-1H in kidney recipients with delayed graft function at the American Transplant Congress 2004 [29]. They compared patients treated with Campath-1H (30 mg given on the day of surgery and day 1) to a contemporaneous control group given various induction agents. All patients were treated with triple therapy maintenance (calcineurin inhibitor, Cellcept and steroids). At 3 months the Campath-1H patients had experienced 12.5% rejection, the anti-CD25 group 35%, the Thymoglobulin induction group 36%, and those categorized as 'other' 61%. Graft survival was significantly better in the Campath-1H group without a difference in infection rate, or patient survival.

The promising early reports using Campath-1H to minimize immunosuppression and potentially induced tolerance has stimulated several other groups to incorporate it in a variety of treatment strategies. The group at the University of Pittsburgh, after their initial experience with a Thymoglobulin-based tolerogenic immunosuppression protocol, has now substituted Campath-1H. The principles of immunosuppression have been outlined in detail by Starzl and colleagues [30]. Preliminary results of their experience with Campath-1H were reported at the American Transplant Congress in 2004. This group has pursued a strategy of pretreatment with Campath-1H, low dose tacrolimus and a progressive weaning of the tacrolimus. Campath-1H (one dose of 30 mg) was given prior to surgery and tacrolimus monotherapy was started on day 1 with early target levels of 10 ng/ml. Ninety-six patients were included in this trial which included cadaveric (59 patients, 61.5%) and living donors (37 patients, 38.5%). Mean follow-up was 4.8 ± 2.3 months. One patient developed rejection before weaning commenced and one patient developed rejection following weaning attempts. None of the patients developed CMV disease (all were treated with ganciclovir prophylaxis), new onset diabetes or post-transplant lymphoprolifer-ative disease. One patient developed a BK virus infection. At the time of this report none of the patients had been completely weaned from tacrolimus. These promising preliminary results suggest that Campath-1H may play a major role in minimizing immunosuppression if an effective strategy can be developed.

Other groups reported preliminary results using Campath-1H in steroid-free protocols at the American Transplant Congress of 2004. Vathsala et al. [31] presented preliminary results of a pilot randomized trial comparing Campath-1H to a standard cyclosporine-based protocol. This study attempts to directly com-pare the protocol previously reported by Calne et al. [22] with standard therapy in a group of Asian patients. Campath-1H (20 mg i.v.) was administered 6 h after surgery and 24 h postoperatively. Cyclosporine was given 72 h after the first Campath-1H dose and targeted blood levels were 90–110 thereafter. Steroids were given prior to each Campath-1H dose only. The standard treatment group received prednisone, Cellcept and cyclosporine with targeted blood levels of 180–225 ng/ml. This group reported the results of 30 renal transplant recipients with 6-month follow-up. Chinese patients comprised 36.7% and Filipinos 16.6% of the population. At the latest follow-up 22.2% of the standard treatment group developed acute rejection compared to 27.8% of the Campath-1H patients. Campath-1H patients developed 16.7% steroid-resistant rejections compared to 11.1% in standard treated patients and 5.6% of Campath-1H patients developed recurrent rejections versus none in the standard group. Treatment failures were reported in 27.8% of Campath-1H versus 11.1% in the standard treated patients. 'Serious adverse events' were reported in 33% of the Campath-1H patients com-pared to 22.2% of the standard group. None of the above findings were statisti-cally significant perhaps owing to the small sample size. Maintenance steroids were required in 22.2% of the Campath-1H patients at the latest reported follow-up. Although this is a preliminary report, the rejection rate appears to be very high by current standards for the Campath-1H and standard group as well. This report appears to echo the experience of Kirk and Knechtle in that rejection is not prevented by lymphocyte depletion and develops at an equivalent if not higher rate with Campath-1H induction.

Potdar and colleagues [34] also reported preliminary results of Campath-1H induction (30 mg i.v. preoperatively) followed by either tacrolimus (targeted trough level 10 ng/ml) or Cellcept (in extended donors or patients with long-cold ischemia times) in 20 cadaveric renal transplant recipients. In addition all patients received prednisone 20 mg/day followed by a weaning schedule of 2.5 mg/week. This group reported one rejection in 20 (5%) cadaveric renal transplants with follow-up between 40 and 240 days after surgery. At the latest follow-up,

11 (55%) patients were receiving tacrolimus monotherapy and 9 (45%) Cellcept alone. Additional follow-up will be needed to determine if the addition of steroids to lymphocyte depletion is safe and capable of inducing any degree of tolerance.

Thymoglobulin

Most studies of lymphocyte depletion related attempts at tolerance induction have been conducted with Campath-1H but one group has used Thymoglobulin. Thymoglobulin is a rabbit polyclonal, anti-thymocyte antibody, which is a potent T cell depleting agent and has been used in clinical transplantation for over 30 years [32]. Unlike Campath-1H, it has broad antibody specificities and recent studies suggest that its many effects may be permissive to cell chimerism [31].

Starzl et al. [33] reported their experiences in 82 recipients of kidney, liver, pancreas or intestinal transplants who were induced with a 5 mg/kg intravenous dose of Thymoglobulin. This protocol was designed based upon two principles: recipient pretreatment and minimum use of immunosuppression after transplantation. To this end, patients underwent progressive dosage reduction in maintenance immunosuppression to the absolute minimum required to prevent progressive allograft destruction. Methylprednisolone (1–2 g i.v.) was given concomitantly to avoid cytokine release and steroids were not planned thereafter. Tacrolimus (targeted levels initially 10 ng/ml) was used as maintenance immunosuppression and attempts at weaning it began at approximately 3 months after surgery in some patients. Fifty kidney transplant recipients were treated under this protocol and 39 underwent tacrolimus dose weaning. Weaning was not attempted in early graft failures or when physician noncompliance with the protocol resulted in the addition of other immunosuppressants. At 14–17 months the dose intervals in the patients were every other day (n = 1, 3%), three times per week (6, 15.4%), two times per week (11, 28.2%), and once per week (7, 17.9%). For all organs, 43 patients were on spaced dosing at the latest follow-up: 6 patients every other day, 11 three times per week, 15 twice per week and 11 once per week.

Shapiro et al. [34] updated the University of Pittsburgh experience in 150 kidney transplant recipients receiving Thymoglobulin pretreatment. One-year graft and patient survival was 97 and 92%. Acute rejection developed in 37% of the patients prior to weaning. Weaning was initiated in 113 of the 150 patients. Of these, 23% developed acute rejection subsequently. Rejection treatment ranged from steroid boluses to antibody therapy with OKT3 or Campath-1H. At the latest follow-up, 94 patients (63%) of the 150 were undergoing spaced weaning; 35 (23.3%) every other day, 29 (19.3) three times per week, 19 (12.7%) twice per week and 11 (7.3%) once per week of tacrolimus. Despite these encouraging results, this group has terminated Thymoglobulin induced in preference to Campath-1H. As with most lymphocyte-depleting protocol reports, this study has a limited follow-up and much longer observation will be required before the risk of chronic allograft nephropathy (CAN) or late acute rejections leading to graft loss can be evaluated.

Conclusions

The recent excitement over steroid-free lymphocyte-depleting protocols aimed at inducing tolerance has ushered in the possibility of a new era in transplantation. Early hopes of generating and sustaining full tolerance have been tempered by the relatively high rate of rejection in many studies and the consistent requirement for some degree of maintenance immunosuppression in all studies. To date, none of the patients in these protocols have become drug free for a sustained period. Most workers in this field have now resigned themselves to the fact that, if tolerance exists at all, it must be partial and relatively unstable. This has resulted in the use of a growing number of terms to describe this state. Calne calls it 'prope tolerance', Knechtle calls it 'metastable' and Starzl has recently described it as 'partial tolerance' [5, 22, 35]. Brent [3] has recently concluded in reference to the work by Calne et al. '... but in the absence of compelling experimental evidence or the total withdrawal of cyclosporin, it is (in my view) premature to conclude that tolerance, or even prope tolerance, was induced'. Starzl and Knechtle have reframed this argument by suggesting that partial tolerance has been present in all long-surviving transplant recipients from the beginning of modern transplantation [5, 34]. Calne [36] has favored a redefinition of tolerance to one which includes an 'operational tolerant state'. He suggests that 'a useful working definition would be long-term functional graft survival in a patient not requiring maintenance immunosuppression' [35]. He further suggests that tolerance may be analogous to happiness and can be complete or partial.

These protocols have demonstrated that immunosuppression can be minimized but the long-term consequences of possibly inadequate immunosuppression must be addressed by a much longer follow-up. Most reported studies do not have one-year follow-up in all patients and none of them have quantitated the frequency and severity of CAN. The high rates of clinical rejection seen in most studies might increase the risk for CAN and subclinical rejection (which has not been evaluated in these studies) and has recently been confirmed as a risk factor for CAN [37]. Eventually these protocols should be tested in randomized studies against standard immunosuppression protocols. Since they remain in relatively early exploratory stages, comparisons against the prevailing standard therapy should await better definition of dosing and timing of immunosuppression and complete evaluation of the minimum safe maintenance dose requirements. Despite the many unanswered questions, these novel treatment strategies may play an important role in advancing the care of transplant recipients.

References

1　United Network of Organ Sharing Website http://www.optn.org/latestData/rptStrat.asp Based on OPTN data as of June 29, 2004.
2　Brent L: A History of Transplantation Immunology. London, Academic, 1997, pp 230–305.
3　Brent LB: Tolerance and its clinical significance: World J Surg 2000;24:787–792.
4　Brent LB: Tolerance: Then and now. Eur Surg Res 2002;34:154–159.
5　Starzl TE, Murase N, Demetris AJ, Trucco M, Abu-Elmagd K, Gray EA, et al: Lessons of organ-induced tolerance learned from historical clinical experience. Transplantation 2004; 77:926–929.
6　Starzl TE, Zinkernagel RM: Transplantation tolerance from a historical perspective. Nat Rev Immunol 2001;1:233–239.
7　Starzl TE, Murase N: Microchimerism, macrochimerism, and tolerance. Clin Transplant 2000;14: 351–354.
8　Cosimi AB, Sachs DH: Mixed chimerism and transplantation tolerance. Transplantation 2004; 77:943–946.
9　Knechtle SJ, Vargo D, Fechner J, Zhai Y, Wang J, Hanaway MJ, et al: FN18-CRM9 immunotoxin promotes tolerance in primate renal allografts. Transplantation 1997;63:1–6.
10　Cannon JA, Longmire WP: Studies of successful skin homografts in the chicken: Description of a method of grafting and its applications: A technique of investigation. Ann Surg 1952;135:60.
11　Billingham RE, Brent L, Medawar PB: Actively acquired tolerance to foreign cells. Nature 1953; 172:603.
12　Personal communication.
13　Calne R: WOFIE hypothesis: Some thoughts on an approach toward allograft tolerance. Transplant Proc 1996;28:1152.
14　Jonker M, Slingerland W, Ossevoort M, Kuhn E, Neville D, Friend P, et al: Induction of kidney graft acceptance by creating a window of opportunity for immunologic engagement (WOFIE) in rhesus monkeys. Transplant Proc 1998;30:2441–2443.
15　Dresske B, Zavazava N, Huang DS, Lin X, Kremer B, Fandrich F: WOFIE augments the immuno-suppressive potency of FK-506. Window of opportunity for immunological engagement. Transpl Immunol 1998;6:243–249.
16　Dresske B, Huang DS, Lin X, Kremer B, Fandrich F: Synergistic impact of "WOFIE" on the immunosuppressive potency of FK 506 in a heterotopic heart transplantation model in the rat. Transplant Proc 1999;31:1561–1562.
17　Dresske B, Zavazava N, Jenisch S, Exner B, Lenz P, El Mokhtari NE, et al: WOFIE synergizes with calcineurin-inhibitor treatment and early steroid withdrawal in kidney transplantation. Transplantation 2003;75:1286–1291.
18　Simpson D: T-cell depleting antibodies: New hope for induction of allograft tolerance in bone marrow transplantation? BioDrugs 2003;17:147–154.
19　Chakrabarti S, Hale G, Waldmann H: Alemtuzumab (Campath-1H) in allogeneic stem cell trans-plantation: Where do we go from here? Transplant Proc 2004;36:1225–1227.
20　Novitzky N, Rubinstein R, Hallett JM, du Toit CE, Thomas VL: Bone marrow transplantation depleted of T cells followed by repletion with incremental doses of donor lymphocytes for relapsing patients with chronic myeloid leukemia: A therapeutic strategy. Transplantation 2000;69:1358–1363.
21　Flynn JM, Byrd JC: Campath-1H monoclonal antibody therapy. Curr Opin Oncol 2000;12: 574–578.
22　Knechtle SJ: Present experience with Campath-1H in organ transplantation and its potential use in pediatric recipients. Pediatr Transplant 2004;8:106–112.

23   Calne R, Friend P, Moffatt S, Bradley A, Hale G, Firth J, et al: Prope tolerance, perioperative campath 1H, and low-dose cyclosporin monotherapy in renal allograft recipients. Lancet 1998;351:1701–1702.

24   Calne R, Moffatt SD, Friend PJ, Jamieson NV, Bradley JA, Hale G, et al: Campath IH allows low-dose cyclosporine monotherapy in 31 cadaveric renal allograft recipients. Transplantation 1999; 68:1613–1616.

25   Watson CJ, Firth J, Bradley J, Smith KG, Jamieson NV, Friend PJ, Hale G, Waldmann H, Bradley A, Calne RY: Campath 1H (Alemtuzumab) in renal transplantation: 5-year comparative follow up. Am J Transplant 2004;4:405.

26   Kirk AD, Hale DA, Mannon RB, Kleiner DE, Hoffmann SC, Kampen RL, et al: Results from a human renal allograft tolerance trial evaluating the humanized CD52-specific monoclonal antibody Alemtuzumab (CAMPATH-1H). Transplantation 2003;76:120–129.

27   Knechtle SJ, Pirsch JD, H Fechne J Jr, Becker BN, Friedl A, Colvin RB, et al: Campath-1H induction plus rapamycin monotherapy for renal transplantation: Results of a pilot study. Am J Transplant 2003;3:722–730.

28   Rao V, Pirsch JD, Becker BN, Knechtle SJ: Sirolimus monotherapy following Campath-1H induction. Transplant Proc 2003;35:S128–S130.

29   Knechtle SJ, Pirsch JD, Voss BJ, Leverson GE, Becker BN, Fernandez LA, Chin LT, Becker YT, Odorico JS, D'Allessandro AM, Sollinger HW: Campath-1h in patients with delayed graft function: Reduced rejection and improved graft survival. Am J Transplant 2004;4:404.

30   Shapiro R, Tan H, Basu A, Khan A, Gray E, Randhawa PS, Murase N, Zeevi A, Demetris AJ, Woodward J, Jordan ML, Ruppert K, Marcos A, Fung JJ, Starzl TE: Campath-1H preconditioning and tacrolimus monotherapy with subsequent weaning in renal transplant recipients. Am J Transplant 2004;4:405.

31   Vathsala A, Ona ET, Tan S, Suresh S, Yio Chan Y, Lou H, Cabanayan Casasola CB, Seldrup J, Calne R: CAMPASI: A pilot randomized controlled trial of the effectiveness of Campath-1H as an induction agent for prevention of graft rejection and preservation of renal function in patients receiving kidney transplants. Am J Transplant 2004;4:406.

32   Mueller TF: Thymoglobulin: An immunologic overview. Curr Opin Organ Transplant 2003;8: 305–312.

33   Starzl TE, Murase N, Abu-Elmagd K, Gray EA, Shapiro R, Eghtesad B, et al: Tolerogenic immunosuppression for organ transplantation. Lancet 2003;361:1502–1510.

34   Shapiro R, Jordan ML, Basu A, Scantlebury V, Potdar S, Tan HP, et al: Kidney transplantation under a tolerogenic regimen of recipient pretreatment and low-dose postoperative immunosuppression with subsequent weaning. Ann Surg 2003;238:520–525.

35   Knechtle SJ, Burlingham WJ: Metastable tolerance in nonhuman primates and humans. Transplantation 2004;77:936–939.

36   Calne RY: Prope tolerance: The future of organ transplantation – From the laboratory to the clinic. Transplantation 2004;77:930–932.

37   Montagnino G, Banfi G, Campise MR, Passerini P, Aroldi A, Cesana BM, Ponticelli C: Impact of chronic allograft nephropathy and subsequent modifications of immunosuppressive therapy on late graft outcomes in renal transplantation. Transplantation 2004;78:242–249.

Jerry McCauley
Transplantation Institute, University of Pittsburgh School of Medicine
Pittsburgh, PA 15215 (USA)
Tel. +1 412 647 2561, Fax +1 412 647 6222, E-Mail jmcc@pitt.edu

Ronco C, Chiaramonte S, Remuzzi G (eds): Kidney Transplantation: Strategies to Prevent Organ Rejection. Contrib Nephrol. Basel, Karger, 2005, vol 146, pp 54–64

....

Chronic Graft Loss

Immunological and Non-Immunological Factors

Maria P. Hernandez-Fuentes, Robert I. Lechler

Department of Immunology, Faculty of Medicine, Imperial College London, London, UK

Abstract

Aims: Late loss of kidney grafts is an ongoing problem in the field of transplantation. This is caused by immunological and non-immunological factors, the main immunological driver of rejection is the immune response against HLA molecules that differ between donor and recipient. *Methods*: To measure the anti-donor responses that a recipient can mount, we have been quantifying anti-donor T-cell frequencies in recipients of renal transplants for several years. Anti-donor direct and indirect pathway frequencies have been measured in vitro in kidney and heart transplant patients by Limiting dilution analysis and other methods. Further, to elucidate the role of CD4+CD25+ regulatory T-cells, these cells have been depleted in ex vivo assays of cellular function. Antigen specific CD4+CD25+ cell lines are being expanded in vitro with a view to using them in immunotherapeutic strategies. *Results*: Frequencies of T-cells with direct pathway anti-donor specificity decline in most patients, while those with indirect anti-donor specificity increase in frequency in patients with late graft failure. In keeping with results from experimental models of transplantation tolerance, evidence for allospecific regulatory cells was found in some patients with good, stable transplant function. Interestingly, the regulatory cells appeared to have indirect allospecificity, and no evidence of direct pathway regulation was observed. *Conclusions*: The indirect pathway anti-donor allore-sponse poses the major threat to long-term transplant survival. Indirect pathway regulatory T-cells arise in some patients. These data are consistent with the hypothesis that tolerance strategies require shrinkage of the direct, and regulation of the indirect, anti-donor response.

Renal transplantation is a successful therapy for end-stage renal failure. With the increase in patients entering the waiting lists and the lack of similar increase in donor availability the long-term success of transplantation is a pressing clinical need. This will help to reduce the number of patients entering the waiting list due to the failure of a first transplant. New immunosuppressive drugs have been very successful in improving short-term allograft survival and there is emerging data on some improvement in long-term survival [1]. Nonetheless, late deterioration of allografts remains an important problem, particularly in view of the increasing demand for transplants. Kidney and hearts allografts currently fail at a rate of 5% each year post-transplantation [2].

Several factors contribute to the pathogenesis of late graft loss. To better understand the mechanisms involved in this process, the terminology is being redefined. Chronic rejection was used in the initial years to describe slow, late deterioration of graft function. Recently, the term is limited to mean late graft loss caused by a host-anti-graft immune response [2–4]. The array of changes found in biopsies of grafts with progressive dysfunction is referred to as chronic allograft nephropathy or CAN. This is characterized by chronic interstitial fibrosis, tubular atrophy, vascular occlusive changes and glomerulopathy [5, 6]. We will review in this chapter, the immunological and non-immunological factors that contribute to chronic allograft nephropathy, the latter having grown in importance lately due to the better control of rejection [3].

Immune Allorecognition of Graft Tissue

Alloantigens in the graft tissue are recognized by T-cells in different forms. Intact allogeneic HLA molecules on the surface of donor tissue are able to directly activate T-cells, this is referred to as the direct pathway of allorecognition [7, 8]. This is a very powerful response, as a high frequency of T-cells are activated in this manner, and it is thought to trigger early graft rejection. Direct pathway T-cell activation is most efficiently achieved by donor bone marrow-derived antigen presenting cells, most importantly, tissue dendritic cells that migrate to draining lymphoid tissue shortly after transplantation. The second pathway of MHC allorecognition is referred to as the 'indirect' pathway and involves the internalization, processing, and presentation of alloantigens as peptides bound to recipient MHC molecules (fig. 1).

The involvement of this latter pathway in transplant rejection was first proposed on the basis of observations in a rat kidney transplant model [9]. Since those early observations, we, and others, have provided evidence that indirect allorecognition is an important driver of late transplant rejection [10–12]. In addition it has been shown that T-cell help for B-cell IgG alloantibody production after skin transplantation in mice depends on the ability to generate an indirect response [13].

Experiments from our group and from other labs have provided substantial evidence, in both rodent and human models, that after the efflux of highly immunogenic antigen presenting cells from a solid organ graft, the strength of

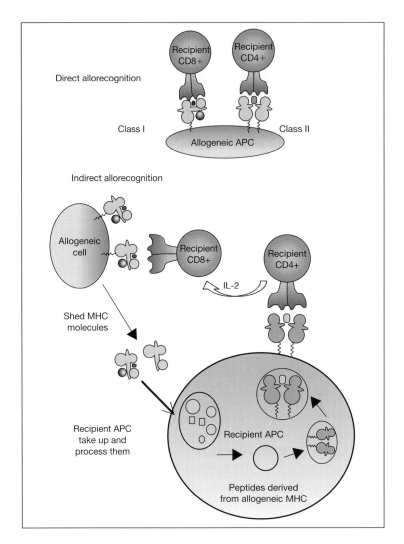

Fig. 1. Direct and indirect pathways of allorecognition. Direct allorecognition: Intact allogeneic MHC molecules being recognized by recipient T-cells. CD8+ T-cells are activated by recognition of class MHC I molecules, whereas CD4+ cells are activated by the recognition of class II molecules. In the indirect pathway allogeneic MHC molecules are shed from the graft and these molecules are taken up and processed by the recipient antigen presenting cells (APCs). Peptides derived from the allogeneic molecules are presented in the context of the appropriate restriction elements. Indirect pathway CD4+ T-cells are able to provide help to direct pathway CD8+ T-cells.

the direct response diminishes with time after transplantation [9, 14–16]. The implication is that then the indirect response becomes predominant [10, 17, 18].

Early Types of Rejection

Hyperacute Rejection (HAR)
Hyperacute rejection is the term applied to very early graft loss, usually within the first 48 h. It occurs when preformed antibodies are present in the recipient's serum, specific for donor antigens expressed on graft vascular endothelial cells. Such antibodies fall into two main categories. Low affinity IgM antibodies, which are specific for ABO blood group antigens, form the first group and they mandate ABO blood matching in solid organ transplantation. Similar antibodies exist in humans against the galactose-α-1–3-galactose epitope present in all other mammals and constitute one of the major impediments to successful xenotransplantation [19]. The second group of antibodies consists of high affinity IgG antibodies directed against HLA class I antigens. These usually occur as a result of previous immunization, by blood transfusions, pregnancies or failed allografts. They also occur in 1% of the population for no obvious reason [20]. The binding of these antibodies triggers activation of clotting, complement and kinin cascades leading to intravascular thrombosis, ischemia and subsequent necrosis. The different pretransplant cross-match techniques (cytotoxic Ab detection, ELISA, FlowPRA) have proved to be successful at virtually eliminating unexpected hyperacute rejection from clinical practice.

Acute Rejection
In the absence of any preformed antibodies, solid organ grafts can still be rapidly rejected after a few days. In the clinical situation, with immunosuppression, this form of rejection usually occurs between 5 days and 3 months after transplantation. The histological findings reveal a diffuse interstitial cellular infiltrate composed of both CD4+ and CD8+ T-cells. This picture is dominated by CD8+ T-cells with an activated or memory, CD45RO+, phenotype [21]. Macrophages are also present to a lesser extent. Initially it was thought that the CD8+ T-cell was foremost in the acute rejection process, as it was demonstrated that they were directly cytotoxic to allogeneic tissue in vitro [22]. There is evidence that donor-specific precursor frequencies rise at the time of acute rejection and drop again when it resolves [23, 24]. CD8+ cytotoxicity is mediated by perforin, granzyme B and Fas-mediated pathways [25–27]. In some animal models it is notable that both CD4+ T-cell and CD8+ T-cell populations can reject solid organ allografts independently, while in others there is some evidence that CD4+ cells are an absolute requirement [28, 29]. In the clinical setting it is

likely that both cell subtypes are involved in the rejection process. In summary, it appears that the acute rejection process is a complex event composed of many effector cells including CD4+ T-cells, CD8+ T-cells and macrophages.

Chronic Rejection

Despite limiting the term 'chronic rejection (CR)' to describe the immune-mediated chronic changes present in late graft dysfunction, there are disagreements about the histological changes that constitute CR. The Banff criteria include extension of interstitial fibrosis, tubular atrophy, mesangial matrix increase, chronic glomerular changes (presence of 'double contours' in capillary loops thought to be secondary to basement membrane duplication) and chronic vascular changes. The vascular features of CR are disruption of the elastic lamina, the presence of inflammatory cells in the intima (endothelialitis) and fibrous intimal thickening due to the proliferation of myofibroblasts [5]. A similar description is given by two recent studies [6, 30]. Some authors claim that the specific changes related to immune-related responses are endothelialitis, tubulitis and complement (C4d) deposition in peritubular capillaries [3, 4]. Others argue that the vascular changes are the primary immunological insult and the parenchymal fibrosis changes are secondary to the ischemia [2]. There is common consent that the detection of C4d deposits in the presence of donor-specific alloantibodies in the circulation implies a B-cell involvement in CR [3, 4, 31]. These findings suggest that antibody-mediated rejection is important [32]. Certainly if anti-donor antibodies are transferred into mice that have a functioning cardiac allograft and no native antibodies, then graft atherosclerosis can be induced [33]. Indeed, it has been shown in human recipients of renal, cardiac and lung allografts that the development of anti-HLA antibodies is linked to the development of CR [34, 35]. Although common consent has not yet been reached concerning the histology of CR, it is clear that there are risk factors linking chronic transplant dysfunction and the anti-donor immune response.

Immunological Factors Influencing Late Graft Loss

First, independently of other factors, good HLA matching has a significant impact on graft survival rates and increases the half-life of kidney grafts [36–38]. In addition, CR is less common in grafts that are better matched, whether they be from live related or cadaveric donors [39]. Secondly, it has been shown that episodes of acute rejection strongly predict the development of CR, implying that the two processes may share common etiologies [37, 40–42].

Even subclinical acute rejection found in protocol biopsies is related to worse long-term kidney function [43]. The fact that increased frequencies of T-cells activated through the indirect pathway are associated to chronic graft failure also suggests involvement of the immune system in this process. Furthermore, the induction of tolerance in this pathway is a requirement for long-term transplant survival in experimental models [44, 45]. Thirdly, alloantibodies can also cause allograft rejection [46, 47]. Their influence on the transplant outcome has long been recognized and they are routinely determined in transplant recipients whose graft function starts to deteriorate. Such antibodies generally recognize the MHC molecules expressed on the graft [46, 48] but anti-endothelial non-MHC antibodies have also been associated with CR [49, 50].

Three effector mechanisms have been recently reviewed implicated in chronic vascular rejection: (1) a CD4+ T-cell-mediated delayed type hypersensitivity response that locally activates macrophages and affects vascular endothelial cells, (2) direct cytolysis of graft parenchymal or vascular cells mediated by CD8+ T-cells and (3) antibody binding to endothelial cells and locally activating the complement system [2].

Non-Immunological Factors Influencing Late Graft Loss

When studying etiological factors involved in late graft loss, non-immunological factors invariably emerge in all of the series. These factors seem to be gaining importance as rejection is being decreased by new immunosuppressive regimens and the demand for organs has led to the use of older donors, kidneys with longer ischemic times, and non-heart beating donors. Donor age has an important negative impact on long-term outcome of kidney transplants that has been recognized for a long time [51]. Other donor characteristics may also be influential, such as size, quality and previous stresses such as hypertension and diabetes [3]. Calcineurin inhibitor nephrotoxicity is also universally recognized as a damaging factor. It is possible to recognize it in biopsies as nodular hyaline deposits in the periphery of arterioles [4–6]. In fact, it has been found in the majority (96.8%) of a recent series of kidney-pancreas transplant biopsies 10 years after transplantation [6]. With the generalized use of these immunosuppressive drugs, it is very difficult to separate the injury secondary to calcineurin inhibitor use from other factors involved in chronic lesions. The nephrotoxicity caused by this family of drugs is amply demonstrated in patients treated with calcineurin inhibitor in the absence of a kidney transplant. Ischemic reperfusion injury at the time of transplantation can induce hypoxia and stimulate the secretion of proinflammatory mediators that can initiate or worsen the immune-mediated injury that develops in

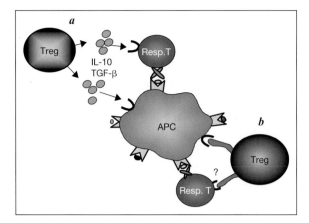

Fig. 2. Mechanisms of action of CD4+ CD25+ T-cells. Regulatory T-cells may suppress through different mechanisms. *a* In vivo models suggest that IL-10 and/or TGF-β play a key role in mediating suppression. *b* While in vitro studies indicate that cell contact is essential, involving as yet undefined molecules.

the graft [2]. Other recipient factors include infection (cytomegalovirus, or polyoma BK virus), hypertension and proteinuria, recurrent original disease and de novo disease [3, 30]. These factors may increase graft injury or promote its irreversibility; thus, it is important to study their effect and to take all possible steps to minimize their negative impact on long-term graft survival.

Role of Regulatory Cells

In the pursuit of clinical transplantation tolerance, the emergence of regulatory T-cells that hold the anti-donor immune response in check appears to be a crucial requirement. These cells express CD25, the α chain of the IL-2 receptor and have been described to suppress the activity of effector CD4+ or CD8+ cells in vitro and in vivo in an antigen nonspecific (bystander) fashion [52–56]. Some of the possible mechanisms of action of these cells are described in figure 2. Depletion of CD4+CD25+ cells prevents the transfer of tolerance by CD4+ T-cells from a transplant-bearing animal [57]. The enrichment and adoptive transfer of these cells has allowed the emergence of transplantation tolerance in murine transplantation models [58, 59]. In vitro assays using depletion of these cells from kidney transplant recipients have shown that they primarily regulate the indirect pathway of allorecognition [60]. Moreover, we have in vitro data to suggest that the immunoregulatory function of these cells is not impaired in the presence of drugs such as rapamycin or anti-CD25 monoclonal antibody – Basiliximab [61]. Evidence from mouse models and human data show that these cells are able to regulate low-frequency responses. It follows that CD4+CD25+ T-cells primarily regulate indirect pathway alloresponses that are usually of very low frequency [18]. The possibility exists for the ex vivo manipulation of these cells to be used as a form of adoptive T-cell therapy to prevent transplant rejection, which could function even in the presence of general immunosuppression. Initial data has been obtained that indicates this may be possible [62].

Conclusion

In order to approach the goal of transplantation tolerance, several key questions remain to be addressed. The role of the indirect pathway of allorecognition in late graft rejection calls for the design of therapeutic strategies that diminish its influence and control these responses. An active investigation is aimed at defining the mechanisms responsible for CR. Interventions that target nonimmunological factors are likely to be necessary, in concert with improved approaches to tolerance induction if the lifespan of transplanted organs is to be enhanced. The evolution of cellular and molecular techniques that help us measure the establishment of tolerance will help to monitor the efficiency of these tolerance-inducing strategies.

References

1   Collaborative-Transplant-Study-Group: CTS-K-15103E-Feb2004. 2004. http://www.ctstransplant.org/public/literature.html
2   Libby P, Pober JS: Chronic rejection. Immunity 2001;14:387–397.
3   Gourishankar S, Halloran PF: Late deterioration of organ transplants: A problem in injury and homeostasis. Curr Opin Immunol 2002;14:576–583.
4   Colvin RB: Chronic allograft nephropathy. N Engl J Med 2003;349:2288–2290.
5   Racusen LC, Solez K, Colvin RB, Bonsib SM, Castro MC, Cavallo T, Croker BP, Demetris AJ, Drachenberg CB, Fogo AB, Furness P, Gaber LW, Gibson IW, Glotz D, Goldberg JC, Grande J, Halloran PF, Hansen HE, Hartley B, Hayry PJ, Hill CM, Hoffman EO, Hunsicker LG, Lindblad AS, Yamaguchi Y, et al: The Banff 97 working classification of renal allograft pathology. Kidney Int 1999;55:713–723.
6   Nankivell BJ, Borrows RJ, Fung CL, O'Connell PJ, Allen RD, Chapman JR: The natural history of chronic allograft nephropathy. N Engl J Med 2003;349:2326–2333.
7   Lombardi G, Barber L, Sidhu S, Batchelor JR, Lechler RI: The specificity of alloreactive T cells is determined by MHC polymorphisms which contact the T cell receptor and which influence peptide binding. Int Immunol 1991;3:769–775.
8   Hernandez-Fuentes MP, Baker RJ, Lechler RI: The alloresponse. Rev Immunogenet 1999;1:282–296.
9   Lechler RI, Batchelor JR: Restoration of immunogenicity to passenger cell-depleted kidney allografts by the addition of donor strain dendritic cells. J Exp Med 1982;155:31–41.

10 Vella JP, Spadafora FM, Murphy B, Alexander SI, Harmon W, Carpenter CB, Sayegh MH: Indirect allorecognition of major histocompatibility complex allopeptides in human renal transplant recipients with chronic graft dysfunction. Transplantation 1997;64:795–800.

11 SivaSai KS, Smith MA, Poindexter NJ, Sundaresan SR, Trulock EP, Lynch JP, Cooper JD, Patterson GA, Mohanakumar T: Indirect recognition of donor HLA class I peptides in lung transplant recipients with bronchiolitis obliterans syndrome. Transplantation 1999;67:1094–1098.

12 Hornick PI, Mason PD, Baker RJ, Hernandez-Fuentes M, Frasca L, Lombardi G, Taylor K, Weng L, Rose ML, Yacoub MH, Batchelor R, Lechler RI: Significant frequencies of T cells with indirect anti-donor specificity in heart graft recipients with chronic rejection. Circulation 2000;101:2405–2410.

13 Steele DJ, Laufer TM, Smiley ST, Ando Y, Grusby MJ, Glimcher LH, Auchincloss H: Two levels of help for B cell alloantibody production. J Exp Med 1996;183:699–703.

14 Mason PD, Robinson CM, Lechler RI: Detection of donor-specific hyporesponsiveness following late failure of human renal allografts. Kidney Int 1996;50:1019–1025.

15 Hornick PI, Mason PD, Yacoub MH, Rose ML, Batchelor R, Lechler RI: Assessment of the contribution that direct allorecognition makes to the progression of chronic cardiac transplant rejection in humans. Circulation 1998;97:1257–1263.

16 Baker RJ, Hernandez-Fuentes MP, Brookes PA, Chaudhry AN, Lechler RI: The role of the allograft in the induction of donor-specific T cell hyporesponsiveness. Transplantation 2001; 72:480–485.

17 Liu Z, Colovai AI, Tugulea S, Reed EF, Fisher PE, Mancini D, Rose EA, Cortesini R, Michler RE, Suciu Foca N: Indirect recognition of donor HLA-DR peptides in organ allograft rejection. J Clin Invest 1996;98:1150–1157.

18 Baker RJ, Hernandez-Fuentes MP, Brookes PA, Chaudhry AN, Cook HT, Lechler RI: Loss of direct and maintenance of indirect alloresponses in renal allograft recipients: Implications for the pathogenesis of chronic allograft nephropathy. J Immunol 2001;167:7199–7206.

19 Parker W, Bruno D, Holzknecht ZE, Platt JL: Characterization and affinity isolation of xenoreactive human natural antibodies. J Immunol 1994;153:3791–3803.

20 Scornik JC, Salomon DR, Howard RJ, Pfaff WW: Evaluation of antibody synthesis in broadly sensitized patients. Transplantation 1988;45:95–100.

21 Ibrahim S, Dawson DV, Sanfilippo F: Predominant infiltration of rejecting human renal allografts with T cells expressing CD8 and CD45RO. Transplantation 1995;59:724–728.

22 Hayry P, Defendi U: Mixed lymphocyte culture produce effector cells: In vitro model for allograft rejection. Science 1970;168:133–135.

23 Ouwehand AJ, Baan CC, Roelen DL, Vaessen LM, Balk AH, Jutte NH, Bos E, Claas FH, Weimar W: The detection of cytotoxic T cells with high-affinity receptors for donor antigens in the transplanted heart as a prognostic factor for graft rejection. Transplantation 1993;56: 1223–1229.

24 Mestre M, Massip E, Bas J, Alsina J, Romeu A, Castelao AM, Buendia E, Grinyo JM: Longitudinal study of the frequency of cytotoxic T cell precursors in kidney allograft recipients. Clin Exp Immunol 1996;104:108–114.

25 Vasconcellos LM, Asher F, Schachter D, Zheng XX, Vasconcellos LH, Shapiro M, Harmon WE, Strom TB: Cytotoxic lymphocyte gene expression in peripheral blood leukocytes correlates with rejecting renal allografts. Transplantation 1998;66:562–566.

26 Sharma VK, Bologa RM, Li B, Xu GP, Lagman M, Hiscock W, Mouradian J, Wang J, Serur D, Rao VK, Suthanthiran M: Molecular executors of cell death – Differential intrarenal expression of Fas ligand, Fas, granzyme B, and perforin during acute and/or chronic rejection of human renal allografts. Transplantation 1996;62:1860–1866.

27 Wever PC, Boonstra JG, Laterveer JC, Hack CE, van der Woude FJ, Daha MR, ten Berge IJ: Mechanisms of lymphocyte-mediated cytotoxicity in acute renal allograft rejection. Transplantation 1998;66:259–264.

28 Krieger NR, Yin DP, Garrison Fathman C: CD4+ but not CD8+ cells are essential for allorejection. J Exp Med 1996;184:2013–2018.

29 Krieger NR, Ito H, Fathman CG: Rat pancreatic islet and skin xenograft survival in CD4 and CD8 knockout mice. J Autoimmun 1997;10:309–315.

30   Grimm PC, Nickerson P, Gough J, McKenna R, Jeffery J, Birk P, Rush DN: Quantitation of allograft fibrosis and chronic allograft nephropathy. Pediatr Transplant 1999;3:257–270.

31   Sijpkens YW, Joosten SA, Wong MC, Dekker FW, Benediktsson H, Bajema IM, Bruijn JA, Paul LC: Immunologic risk factors and glomerular C4d deposits in chronic transplant glomerulopathy. Kidney Int 2004;65:2409–2418.

32   Racusen LC, Colvin RB, Solez K, Mihatsch MJ, Halloran PF, Campbell PM, Cecka MJ, Cosyns JP, Demetris AJ, Fishbein MC, Fogo A, Furness P, Gibson IW, Glotz D, Hayry P, Hunsickern L, Kashgarian M, Kerman R, Magil AJ, Montgomery R, Morozumi K, Nickeleit V, Randhawa P, Regele H, Seron D, Seshan S, Sund S, Trpkov K: Antibody-mediated rejection criteria – An addition to the Banff 97 classification of renal allograft rejection. Am J Transplant 2003;3:708–714.

33   Russell PS, Chase CM, Winn HJ, Colvin RB: Coronary atherosclerosis in transplanted mouse hearts. II. Importance of humoral immunity. J Immunol 1994;152:5135–5141.

34   Suciu Foca N, Reed E, Marboe C, Harris P, Yu PX, Sun YK, Ho E, Rose E, Reemtsma K, King DW: The role of anti-HLA antibodies in heart transplantation. Transplantation 1991;51:716–724.

35   Sundaresan S, Mohanakumar T, Smith MA, Trulock EP, Lynch J, Phelan D, Cooper JD, Patterson GA: HLA-A locus mismatches and development of antibodies to HLA after lung transplantation correlate with the development of bronchiolitis obliterans syndrome. Transplantation 1998;65: 648–653.

36   Takemoto SK, Terasaki PI, Gjertson DW, Cecka JM: Twelve years' experience with national sharing of HLA-matched cadaveric kidneys for transplantation. N Engl J Med 2000;343:1078–1084.

37   Opelz G, Wujciak T, Dohler B, Scherer S, Mytilineos J: HLA compatibility and organ transplant survival. Collaborative Transplant Study. Rev Immunogenet 1999;1:334–342.

38   Collaborative Transplant Study Group: First cadaver kidney transplants 1985–2002, HLA-+A+B+DR mismatches. 2004. http://www.ctstransplant.org/public/data/html_all/CTS-K-21103-Feb2004.html

39   Cecka M: Clinical outcome of renal transplantation. Factors influencing patient and graft survival. Surg Clin North Am 1998;78:133–148.

40   Almond PS, Matas A, Gillingham K, Dunn DL, Payne WD, Gores P, Gruessner R, Najarian JS: Risk factors for chronic rejection in renal allograft recipients. Transplantation 1993;55:752–756.

41   Tesi RJ, Elkhammas EA, Henry ML, Davies EA, Salazar A, Ferguson RM: Acute rejection episodes: Best predictor of long-term primary cadaveric renal transplant survival. Transplant Proc 1993;25:901–902.

42   Matas AJ: Impact of acute rejection on development of chronic rejection in pediatric renal transplant recipients. Pediatr Transplant 2000;4:92–99.

43   Shishido S, Asanuma H, Nakai H, Mori Y, Satoh H, Kamimaki I, Hataya H, Ikeda M, Honda M, Hasegawa A: The impact of repeated subclinical acute rejection on the progression of chronic allograft nephropathy. J Am Soc Nephrol 2003;14:1046–1052.

44   Azuma H, Chandraker A, Nadeau K, Hancock WW, Carpenter CB, Tilney NL, Sayegh MH: Blockade of T-cell costimulation prevents development of experimental chronic renal allograft rejection. Proc Natl Acad Sci USA 1996;93:12439–12444.

45   Yamada A, Chandraker A, Laufer TM, Gerth AJ, Sayegh MH, Auchincloss H Jr: Recipient MHC class II expression is required to achieve long-term survival of murine cardiac allografts after costimulatory blockade. J Immunol 2001;167:5522–5526.

46   Pelletier RP, Hennessy PK, Adams PW, VanBuskirk AM, Ferguson RM, Orosz CG: Clinical significance of MHC-reactive alloantibodies that develop after kidney or kidney-pancreas transplantation. Am J Transplant 2002;2:134–141.

47   Vongwiwatana A, Tasanarong A, Hidalgo LG, Halloran PF: The role of B cells and alloantibody in the host response to human organ allografts. Immunol Rev 2003;196:197–218.

48   van Kampen CA, Roelen DL, Versteeg-van der Voort Maarschalk MF, Hoitsma AJ, Allebes WA, Claas FH: Activated HLA class I-reactive cytotoxic T lymphocytes associated with a positive historical crossmatch predict early graft failure. Transplantation 2002;74:1114–1119.

49   Wu GD, Jin YS, Salazar R, Dai WD, Barteneva N, Barr ML, Barsky LW, Starnes VA, Cramer DV: Vascular endothelial cell apoptosis induced by anti-donor non-MHC antibodies: A possible injury pathway contributing to chronic allograft rejection. J Heart Lung Transplant 2002;21: 1174–1187.

50 Magro CM, Klinger DM, Adams PW, Orosz CG, Pope-Harman AL, Waldman WJ, Knight D, Ross P Jr: Evidence that humoral allograft rejection in lung transplant patients is not histocompatibility antigen-related. Am J Transplant 2003;3:1264–1272.

51 Opelz G: Factors influencing long-term graft loss. The Collaborative Transplant Study. Transplant Proc 2000;32:647–649.

52 Sakaguchi S, Sakaguchi N, Asano M, Itoh M, Toda M: Immunologic self-tolerance maintained by activated T cells expressing IL-2 receptor alpha-chains (CD25). Breakdown of a single mechanism of self-tolerance causes various autoimmune diseases. J Immunol 1995;155: 1151–1164.

53 Thornton AM, Shevach EM: CD4+CD25+ immunoregulatory T cells suppress polyclonal T cell activation in vitro by inhibiting interleukin 2 production. J Exp Med 1998;188:287–296.

54 Lin CY, Graca L, Cobbold SP, Waldmann H: Dominant transplantation tolerance impairs CD8+ T cell function but not expansion. Nat Immunol 2002;3:1208–1213.

55 van Maurik A, Herber M, Wood KJ, Jones ND: Cutting edge: CD4+CD25+ alloantigen-specific immunoregulatory cells that can prevent CD8+ T cell-mediated graft rejection: Implications for anti-CD154 immunotherapy. J Immunol 2002;169:5401–5404.

56 Camara NO, Sebille F, Lechler RI: Human CD4+CD25+ regulatory cells have marked and sustained effects on CD8+ T cell activation. Eur J Immunol 2003;33:3473–3483.

57 Hall BM, Pearce NW, Gurley KE, Dorsch SE: Specific unresponsiveness in rats with prolonged cardiac allograft survival after treatment with cyclosporine. III. Further characterization of the CD4+ suppressor cell and its mechanisms of action. J Exp Med 1990;171:141–157.

58 Kingsley CI, Karim M, Bushell AR, Wood KJ: CD25+CD4+ regulatory T cells prevent graft rejection: CTLA-4- and IL-10-dependent immunoregulation of alloresponses. J Immunol 2002;168:1080–1086.

59 Hara M, Kingsley CI, Niimi M, Read S, Turvey SE, Bushell AR, Morris PJ, Powrie F, Wood KJ: IL-10 is required for regulatory T cells to mediate tolerance to alloantigens in vivo. J Immunol 2001;166:3789–3796.

60 Salama AD, Najafian N, Clarkson MR, Harmon WE, Sayegh MH: Regulatory CD25+ T cells in human kidney transplant recipients. J Am Soc Nephrol 2003;14:1643–1651.

61 Game DS, Hernandez-Fuentes MP, Lechler RI: Everollmus and Basiliximab permit suppression by human CD4+CD25+ cell in vitro. Am J Transplant 2004; in press.

62 Jiang S, Camara N, Lombardi G, Lechler RI: Induction of allopeptide-specific human CD4+CD25+ regulatory T cells ex vivo. Blood 2003;102:2180–2186. Epub 2003 May 29.

Dr. Maria Hernandez Fuentes
Department of Immunology, Faculty of Medicine
Imperial College London, Hammersmith Campus
Du Cane Road, London W12 0NN (UK)
Tel. +44 (0) 20 8383 1714, Fax +44 (0) 20 8383 2788
E-Mail m.hernandezfuente@imperial.ac.uk

Ronco C, Chiaramonte S, Remuzzi G (eds): Kidney Transplantation: Strategies to Prevent
Organ Rejection. Contrib Nephrol. Basel, Karger, 2005, vol 146, pp 65–72

······················

Lymphocyte Depletion as a Barrier to Immunological Tolerance

David Neujahr, Laurence A. Turka

University of Pennsylvania, Philadelphia, Pa., USA

Abstract

Lymphocyte depletion is a commonly used approach in clinical renal transplantation as part of standard induction immunosuppression therapy. It is also increasing being incorporated into new approaches aimed at inducing transplantation tolerance. While the theoretical basis for eliminating alloreactive T cells as a means to prevent rejection and induce tolerance is sound, new evidence suggests that global non-specific deletion may, by promoting the development of memory T cells, actually present a barrier to tolerance induction.

The Case against Memory T Cells

B cell memory has long been known to have a negative impact in clinical transplantation [1]. In some cases this can be predicted by screening potential recipients for preformed antibodies directed against allo-MHC. In other cases pretransplant B cell immunity only becomes apparent in the post-transplant setting with the rapid induction of antibody by B cells and plasma cells. The case against T cells has been long assumed, but has been harder to prove due in part to the lack of a rapid and effective screening test for memory T cells. Evidence implicating memory T cells derives largely from the experimental setting, and includes the fact that presensitized animals have increased numbers of antigen reactive cells [2]. Further, animals primed with allogenic skin grafts rapidly reject subsequent cardiac allografts from the sensitizing strain even in the setting of normally graft-prolonging therapies such as costimulatory blockade. Adoptive transfer experiments elegantly demonstrate that this rejection is mediated by memory T cells [3, 4]. Finally, in clinical transplantation, patients who receive an organ for which they have been previously sensitized, but who are nevertheless serologically 'cross-match negative' still have statistically worse survival when compared with unsensitized recipients [5, 6].

The difficulty in achieving stable acceptance of allografts in humans or presensitized animals is largely ascribed to the robust nature of memory T cells when compared with naïve cells. Indeed, on a per-cell basis, memory T cells show more exuberant proliferation, greater production of cytokines such as IFN-γ, have increased lytic capacity, and do not require CD28 for costimulation [7–11]. This enhanced ability to respond to alloantigen and decreased need for costimulation may explain the discrepancy between the ease of tolerizing unsensitized animals maintained in a pathogen-free environment and the difficulty in tolerizing outbred mice, primates and humans who have generated a lifetime of memory cells. This immunological history leads to an immune repertoire whereby 40–50% of T cells in the peripheral blood of adult humans and non-human primates have a surface phenotype consistent with memory [12, 13]. Of course, even in immunologically naïve animals, the responder frequency of T cells directed against allo-MHC is quite high. For example, in MHC-mismatched mice several studies estimate the responder frequency to be 7–8% of all circulating T cells [14, 15]. Thus the added burden of memory compounded with a high frequency of alloreactivity becomes a very difficult problem to surmount.

To address the problem of alloreactive T cells, protocols in humans have been developed to effect transient depletion of T cells. These protocols include the use of agents such as anti-thymocyte globulin, OKT3, daclizumab, and Campath 1H [16–18]. Additionally, depletion of a subset of T cells has been accomplished in mice using antibodies directed against CD45RB [19, 20]. In each case, large-scale removal of all, or a portion of circulating T cells has been accomplished. In many trials depletion has led to prolonged allograft survival [17, 21–23]. Unfortunately in no situation to date using human subjects has large-scale depletion led to a state of donor-specific tolerance (unless depletion itself is merely part of a conditioning regimen for bone marrow transplantation to achieve mixed chimerism). Here we explore the immunological aftermath following lymphocyte depletion and propose a model by which depletion could prevent tolerance induction.

Lymphopenia and Homeostatic Proliferation

Homeostatic proliferation refers to the division of peripheral T cells in the absence of exogenous stimuli. While originally described in TCR transgenic mice that had undergone immune ablation, homeostatic proliferation is likely to be relevant in humans in situations of transient lymphopenia such as viral infections, or simply as a means of maintaining the peripheral T cell compartment following thymic involution [24–26]. Homeostatic proliferation occurs in all peripheral T cell subsets including CD4, CD8, NKT cells and CD4+CD25+

regulatory T cells (Tregs) [27, 28]. Moreover, homeostatic proliferation is a feature of both naïve and antigen-experienced T cells.

While the precise mechanisms underlying homeostatic proliferation remain to be elucidated, there are several necessary factors. First, proliferation of naïve cells requires contact with self-MHC, as homeostatic proliferation does not occur for CD8 or CD4 cells in animals absent in MHC class I or class II, respectively [24, 29, 30]. In contrast, memory CD8 cells can proliferate in MHC-deficient hosts [31]. Second, homeostatic proliferation requires cytokine stimulation. Experiments utilizing either transgenic mice or knockout mice show that naïve CD4 and CD8 T cells have an absolute requirement for CCL21 and IL-7, respectively [32–34]. With memory CD8 cells, IL-15 can substitute for IL-7. Recent experiments have shown that the STAT5 signaling pathway downstream of cytokine receptors is a crucial regulator of homeostatic proliferation for CD8 T cells. T cells in transgenic mice that constitutively express STAT5b could proliferate in the absence of IL-7 and IL-15 [35]. These studies suggest a paradigm for homeostatic proliferation whereby cytokines and chemokines, which are generated by nonlymphoid epithelial or stromal cells, exist at limiting concentrations within the host. During periods of lymphodepletion, there is less competition for the scarce resource, and hence proliferation occurs. Weak interactions with self-MHC with/without self-peptide provide TCR stimulation akin to that delivered during the process of positive thymic selection.

Homeostatic Proliferation and a Memory Phenotype

Homeostatic proliferation appears to do more than maintain the status quo of the peripheral T cell compartment. Notably, several studies have demonstrated that cells undergoing homeostatic proliferation acquire several features of antigen-experienced memory cells, without known exposure to antigen [9, 36–38]. These attributes include both the appearance of phenotypic surface markers typically associated with memory cells as well as the acquisition of memory function. Homeostatically proliferating cells show increased surface expression of CD44, CD122, and Ly6C, while simultaneously having decreased expression of CD62L [9, 36–38]. However, there are also features that distinguish homeostatically proliferating cells from antigen-driven proliferation. For example, increased expression of early markers of activation such as CD69, CD71 and CD25 (IL2αR), or of alternative costimulatory molecules such as 41-BB has not been observed [9, 36–39]. We have observed that proliferation does lead to increased expression of other novel costimulatory molecules such as ICOS, and CD134 (OX-40), but that the expression of these molecules is decreased in magnitude, and retarded in kinetics compared with bonafide allo-MHC responses [unpubl. observations]. Hence, the kinetics of acquisition of this 'memory-like' phenotype requires more time and/or more rounds of cell division.

T cells undergoing homeostatic proliferation acquire functional properties of memory cells in that they produce greater levels of cytokines, depend less on classical costimulatory interactions through CD28 and CD40L, and are hypersensitive to antigen encounter. While these responses are slightly less dramatic than true antigen experienced T cells, they are markedly increased compared with the responses of naïve T cells [36–38]. One feature of homeostatic proliferation that remains unanswered (and which is critical for developing clinical tolerance protocols) is whether the 'memory-like' phenotype represents an irreversible fate, or if such cells can revert back to naïve cells once proliferation ceases.

Depletion, Proliferation, and Transplant Tolerance

The finding that lymphopenia-induced proliferation led to a memory phenotype prompted us to examine whether or not such an event would impair tolerance induction in the setting of lymphodepletion. To address this question, we studied a mouse model where extensive, but not complete T cell depletion was created. Mice received depleting antibodies against CD4 and CD8 cells to reduce the peripheral T cell compartment by 90–95%. Mice subsequently received an MHC mismatched heart, as well as costimulatory blockade with CTLA4Ig plus donor antigen in the form of irradiated donor splenocytes. This protocol reliably induces stable tolerance in nonlymphopenic recipients. In contrast, transplantation during periods of homeostatic proliferation consistently led to graft rejection [40]. Perhaps more striking was the finding that T cells which had undergone homeostatic proliferation in either SCID mice or antibody-depleted mice, and returned to rest, could then confer dominant resistance to tolerance induction upon transfer into syngeneic, otherwise naive, animals. These experiments suggest that large-scale T cell depletion can result in a significant population of memory or 'memory-like' cells which resist tolerance induction.

The mechanism underlying this resistance is still unknown and at least two nonmutually exclusive mechanisms could explain it. First, naïve cells, effector/memory cells, and regulatory cells could have differential susceptibility to depletion resulting in a relative skewing of the populations prior to proliferation. Second, these different subsets may have variable efficiency with which they proliferate, again perturbing the relative balance of memory and naïve cells (see fig. 1).

Regarding the Treg population, defined as Foxp3-expressing CD4+CD25+ cells, several studies have shown that these cells do indeed

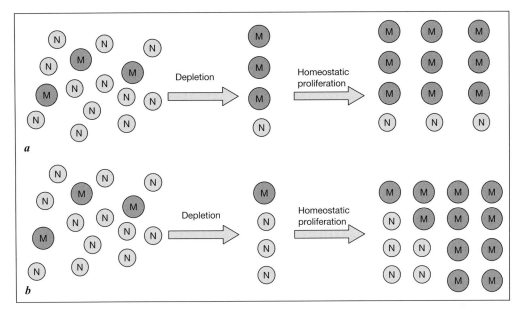

Fig. 1. Two potential mechanisms by which homeostatic proliferation leads to an increased burden of memory T cells. (*a*) Memory T cells (M) are more resistant to depletion than naïve T cells (N). In the immediate aftermath of depletion the residual T cell pool is skewed toward memory cells. Both memory and naïve cells proliferate with equivalent efficiency. (*b*) Memory T cells and naïve T cells are depleted to the same degree. Following depletion, memory cells have greater proliferative capacity and hence increase their relative frequency.

undergo homeostatic proliferation [27, 41]. Following homeostatic proliferation Tregs maintain their regulatory/suppressive capability [39, 40]. However, the efficiency of proliferation relative to other cell populations remains unknown. Further, data in the literature offers contradictory results regarding the ability of Tregs to suppress the proliferation of nonregulatory cells [39, 42].

The murine studies cited above do not imply that lymphodepletion and tolerance induction are mutually exclusive in all cases. For example, the successful use of mixed hematopoietic chimerism in primates and humans shows this is not the case [43–45]. In these protocols, T cell depletion is used to eliminate allo-reactive T cells and 'make space' for bone marrow engraftment. Because donor hematopoietic stem cells are transferred in these protocols, donor-derived dendritic cells can populate the recipient thymus and thereby limit export of donor-reactive T cells [13]. In the case of tolerance to vascularized grafts where macrochimerism is not achieved, the critical issue may be whether or not T cell depletion is complete. Another aspect where clinical transplant tolerance could differ from murine protocols is that lymphodepletion, as in the

Campath 1H trials, leads to profound and long-lasting lymphopenia. In spite of the aggressive nature of lymphodepletion, all of the patients in this protocol experienced episodes of reversible rejection [17]. Conversely, in murine models, large-scale antibody depletion does not result in long-term lymphopenia, with peripheral T cell pools returning to normal within 12 weeks.

Unanswered Questions and Future Directions

Since Medawar's classic demonstration of tolerance using donor marrow transfusion into a newborn mouse 50 years ago, the field of clinical transplantation has made steady advances. Unfortunately, true donor-specific tolerance has been achieved in only a small fraction of patients, with the remainder relegated to the complications of immunosuppressive drugs and infections. A large body of work suggests that preferentially deleting memory cells is a key step toward tolerance. At the same time, accomplishing this task through wholesale T cell depletion may be counter-productive. Perhaps the best scenario is one in which only the allo-reactive population of memory T cells is targeted. While this clearly cannot be achieved based on TCR specificity, the use of surrogate markers, such as cytokine receptors, may be promising [46].

The current paradigm for peripheral T cell tolerance is one in which the mechanisms of deletion and regulation have complementary roles [47]. As we have mentioned, large-scale depletion of the entire T cell pool may be counter-productive. What remains to be seen is whether more specific depleting regimens can be developed, or additional measures can be taken to limit the allo-responsive T cell pool, while not hindering the Treg population. In this regard, a recent report in a stringent model of primate diabetes is of particular interest [46]. These investigators were able to delete alloaggressive lymphocytes, but not CD4+CD25+ regulatory T cells using lytic fusion proteins against IL-2 and IL-15. Hence, one key to the success of depletion protocols may be to understand the basis for regulatory cell sparing and exploiting it.

References

1 Turka LA, Goguen JE, Gagne JE, Milford EL: Presensitization and the renal allograft recipient. Transplantation 1989;47:234–240.
2 Heeger PS, Greenspan NS, Kuhlenschmidt S, Dejelo C, Hricik DE, Schulak JA, Tary-Lehmnn M: Pretransplant frequency of donor-specific, IFN-gamma-producing lymphocytes is a manifestation of immunologic memory and correlates with the risk of posttransplant rejection episodes. J Immunol 1999;163:2267–2275.
3 Valujskikh A, Pantenburg B, Heeger PS: Primed allospecific T cells prevent the effects of costimulatory blockade on prolonged cardiac allograft survival in mice. Am J Transplant 2002;2:501–509.

4   Zhai Y, Meng L, Gao F, Busuttil RW, Kupiec-Weglinski JW: Allograft rejection by primed/memory CD8+ T cells is CD154 blockade resistant: Therapeutic implications for sensitized transplant recipients. J Immunol 2002;169:4667–4673.

5   van Kampen CA, Versteeg-van der Voort Maarschalk MF, Roelen DL, Claas FH: Primed CTLs specific for HLA class I may still be present in sensitized patients when anti-HLA antibodies have disappeared: Relevance for donor selection. Transplantation 2002;73:1286–1290.

6   Worthington JE, Martin S, Al-Husseini DM, Dyer PA, Johnson RW: Posttransplantation production of donor HLA-specific antibodies as a predictor of renal transplant outcome. Transplantation 2003;75:1034–1040.

7   Rogers PR, Dubey C, Swain SL: Qualitative changes accompany memory T cell generation: Faster, more effective responses at lower doses of antigen. J Immunol 2000;164:2338–2346.

8   Veiga-Fernandes H, Walter U, Bourgeois C, McLean A, Rocha B: Response of naive and memory CD8+ T cells to antigen stimulation in vivo. Nat Immunol 2000;1:47–53.

9   Gudmundsdottir H, Turka LA: A closer look at homeostatic proliferation of CD4+ T cells: Costimulatory requirements and role in memory formation. J Immunol 2001;167:3699–3707.

10   Garcia S, DiSanto J, Stockinger B: Following the development of a CD4 T cell response in vivo: From activation to memory formation. Immunity 1999;11:163–171.

11   Bachmann MF, Barner M, Viola A, Kopf M: Distinct kinetics of cytokine production and cytolysis in effector and memory T cells after viral infection. Eur J Immunol 1999;29:291–299.

12   Pitcher CJ, Hagen SI, Walker JM, Lum R, Mitchell BL, Maino VC, et al: Development and homeostasis of T cell memory in rhesus macaque. J Immunol 2002;168:29–43.

13   Adams AB, Pearson TC, Larsen CP: Heterologous immunity: An overlooked barrier to tolerance. Immunol Rev 2003;196:147–160.

14   Suchin EJ, Langmuir PB, Palmer E, Sayegh MH, Wells AD, Turka LA: Quantifying the frequency of alloreactive T cells in vivo: New answers to an old question. J Immunol 2001;166:973–981.

15   Kishimoto K, Sandner S, Imitola J, Sho M, Li Y, Langmuir PB, et al: Th1 cytokines, programmed cell death, and alloreactive T cell clone size in transplant tolerance. J Clin Invest 2002;109:1471–1479.

16   Szczech LA, Berlin JA, Feldman HI: The effect of antilymphocyte induction therapy on renal allograft survival. A meta-analysis of individual patient-level data. Anti-Lymphocyte Antibody Induction Therapy Study Group. Ann Intern Med 1998;128:817–826.

17   Kirk AD, Hale DA, Mannon RB, Kleiner DE, Hoffmann SC, Kampen RL, et al: Results from a human renal allograft tolerance trial evaluating the humanized CD52-specific monoclonal antibody alemtuzumab (CAMPATH-1H). Transplantation 2003;76:120–129.

18   Kuypers DR, Vanrenterghem YF: Monoclonal antibodies in renal transplantation: Old and new. Nephrol Dial Transplant 2004;19:297–300.

19   Rothstein DM, Livak MF, Kishimoto K, Ariyan C, Qian HY, Fecteau S, et al: Targeting signal 1 through CD45RB synergizes with CD40 ligand blockade and promotes long term engraftment and tolerance in stringent transplant models. J Immunol 2001;166:322–329.

20   Molano RD, Pileggi A, Berney T, Poggioli R, Zahr E, Oliver R, et al: Prolonged islet allograft survival in diabetic NOD mice by targeting CD45RB and CD154. Diabetes 2003;52:957–964.

21   Charpentier B, Rostaing L, Berthoux F, Lang P, Civati G, Touraine JL, et al: A three-arm study comparing immediate tacrolimus therapy with antithymocyte globulin induction therapy followed by tacrolimus or cyclosporine A in adult renal transplant recipients. Transplantation 2003;75:844–851.

22   Agha IA, Rueda J, Alvarez A, Singer GG, Miller BW, Flavin K, et al: Short course induction immunosuppression with thymoglobulin for renal transplant recipients. Transplantation 2002;73:473–475.

23   Knechtle SJ, Pirsch JD, H. Fechner JJ, Becker BN, Friedl A, Colvin RB, et al: Campath-1H induction plus rapamycin monotherapy for renal transplantation: Results of a pilot study. Am J Transplant 2003;3:722–730.

24   Kieper WC, Jameson SC: Homeostatic expansion and phenotypic conversion of naive T cells in response to self peptide/MHC ligands. Proc Natl Acad Sci USA 1999;96:13306–13311.

25   Mackall CL, Bare CV, Granger LA, Sharrow SO, Titus JA, Gress RE: Thymic-independent T cell regeneration occurs via antigen-driven expansion of peripheral T cells resulting in a repertoire that is limited in diversity and prone to skewing. J Immunol 1996;156:4609–4616.

26   Zimmerman C, Brduscha-Riem K, Blaser C, Zinkernagel RM, Pircher H: Visualization, characterization, and turnover of CD8+ memory T cells in virus-infected hosts. J Exp Med 1996;183:1367–1375.

27 Gavin MA, Clarke SR, Negrou E, Gallegos A, Rudensky A: Homeostasis and anergy of CD4(+)CD25(+) suppressor T cells in vivo. Nat Immunol 2002;3:33–41.

28 Matsuda JL, Gapin L, Sidobre S, Kieper WC, Tan JT, Ceredig R, et al: Homeostasis of V alpha 14i NKT cells. Nat Immunol 2002;3:966–974.

29 Ernst B, Lee DS, Chang JM, Sprent J, Surh CD: The peptide ligands mediating positive selection in the thymus control T cell survival and homeostatic proliferation in the periphery. Immunity 1999;11:173–181.

30 Muranski P, Chmielowski B, Ignatowicz L: Mature CD4+ T cells perceive a positively selecting class II MHC/peptide complex in the periphery. J Immunol 2000;164:3087–3094.

31 Murali-Krishna K, Lau LL, Sambhara S, Lemonnier F, Altman J, Ahmed R: Persistence of memory CD8 T cells in MHC class I-deficient mice. Science 1999;286:1377–1381.

32 Ploix C, Lo D, Carson MJ: A ligand for the chemokine receptor CCR7 can influence the homeostatic proliferation of CD4 T cells and progression of autoimmunity. J Immunol 2001;167:6724–6730.

33 Tan JT, Ernst B, Kieper WC, LeRoy E, Sprent J, Surh CD: Interleukin (IL)-15 and IL-7 jointly regulate homeostatic proliferation of memory phenotype CD8+ cells but are not required for memory phenotype CD4+ cells. J Exp Med 2002;195:1523–1532.

34 Goldrath AW, Sivakumar PV, Glaccum M, Kennedy MK, Bevan MJ, Benoist C, et al: Cytokine requirements for acute and basal homeostatic proliferation of naive and memory CD8+ T cells. J Exp Med 2002;195:1515–1522.

35 Burchill MA, Goetz CA, Prlic M, O'Neil JJ, Harmon IR, Bensinger SJ, et al: Distinct effects of STAT5 activation on CD4+ and CD8+ T cell homeostasis: Development of CD4+CD25+ regulatory T cells versus CD8+ memory T cells. J Immunol 2003;171:5853–5864.

36 Cho BK, Rao VP, Ge Q, Eisen HN, Chen J: Homeostasis-stimulated proliferation drives naive T cells to differentiate directly into memory T cells. J Exp Med 2000;192:549–556.

37 Murali-Krishna K, Ahmed R: Cutting edge: Naive T cells masquerading as memory cells. J Immunol 2000;165:1733–1737.

38 Goldrath AW, Bogatzki LY, Bevan MJ: Naive T cells transiently acquire a memory-like phenotype during homeostasis-driven proliferation. J Exp Med 2000;192:557–564.

39 Prlic M, Blazar BR, Khoruts A, Zell T, Jameson SC: Homeostatic expansion occurs independently of costimulatory signals. J Immunol 2001;167:5664–5668.

40 Wu Z, Bensinger SJ, Zhang J, Chen C, Yuan X, Huang X, et al: Homeostatic proliferation is a barrier to transplantation tolerance. Nat Med 2004;10:87–92.

41 Cozzo C, Larkin J 3rd, Caton AJ: Cutting edge: Self-peptides drive the peripheral expansion of CD4+CD25+ regulatory T cells. J Immunol 2003;171:5678–5682.

42 Takeda I, Ine S, Killeen N, Ndhlovu LC, Murata K, Satomi S, et al: Distinct roles for the OX40-OX40 ligand interaction in regulatory and nonregulatory T cells. J Immunol 2004;172:3580–3589.

43 Kawai T, Cosimi AB, Colvin RB, Powelson J, Eason J, Kozlowski T, et al: Mixed allogeneic chimerism and renal allograft tolerance in cynomolgus monkeys. Transplantation 1995;59:256–262.

44 Spitzer TR, Delmonico F, Tolkoff-Rubin N, McAfee S, Sackstein R, Saidman S, et al: Combined histocompatibility leukocyte antigen-matched donor bone marrow and renal transplantation for multiple myeloma with end stage renal disease: The induction of allograft tolerance through mixed lymphohematopoietic chimerism. Transplantation 1999;68:480–484.

45 Buhler LH, Spitzer TR, Sykes M, Sachs DH, Delmonico FL, Tolkoff-Rubin N, et al: Induction of kidney allograft tolerance after transient lymphohematopoietic chimerism in patients with multiple myeloma and end-stage renal disease. Transplantation 2002;74:1405–1409.

46 Zheng XX, Sanchez-Fueyo A, Sho M, Domenig C, Sayegh MH, Strom TB: Favorably tipping the balance between cytopathic and regulatory T cells to create transplantation tolerance. Immunity 2003;19:503–514.

47 Lechler RI, Garden OA, Turka LA: The complementary roles of deletion and regulation in transplantation tolerance. Nat Rev Immunol 2003;3:147–158.

Laurence A. Turka
University of Pennsylvania, 700 Clinical Research Building
415 Curie Boulevard, Philadelphia, PA 19104 (USA)
Tel. +1 215 898 1018, Fax +1 215 573 2880, E-Mail turka@mail.med.upenn.edu

Ronco C, Chiaramonte S, Remuzzi G (eds): Kidney Transplantation: Strategies to Prevent Organ Rejection. Contrib Nephrol. Basel, Karger, 2005, vol 146, pp 73–86

....

Monitoring of Immunosuppressive Therapy in Renal Transplanted Patients

S. Chiaramonte, D. Dissegna, C. Ronco

Department of Nephrology, Dialysis and Transplantation,
St. Bortolo Hospital, Vicenza, Italy

Abstract

The regulation of the immunosuppressive therapy after kidney transplantation is the most complex aspect of the management of transplanted patients. Every day the transplant clinician is challenged by need to provide a sufficient immunosuppression to avoid or reduce the risk of rejection without exposing the patient to the risk of developing opportunistic infections or malignancy or toxic side effects. The safety and efficacy profile of immunosuppressive therapy is limited within a narrow therapeutic window whose borders are represented by two clinical conditions such as rejection and drug-related toxicity. The availability of several different drugs allows the clinicians to make multiple choices to individualize treatments according to the specific needs of a single patient. Pharmacokinetic monitoring of the immunosuppressive drugs is an important element in the management of these patients but cannot be considered as the unique driving factor and must be integrated with a careful surveillance and evaluation of all drug-related side effects.

Over the last decade, several new immunosuppressive medications have become available for the maintenance therapy of transplanted patients giving nephrologists the opportunity to optimize the immunosuppressive protocols.

Twenty-five years ago, azathioprine and steroids were considered to be the standard therapy. The availability of calcineurin inhibitors has fundamentally altered immunosuppressive regimens. The addition of mycophenolate mofetil (MMF) and, more recently, rapamycin has further improved the immunosuppressant armamentarium.

Following the early induction therapy, in the long-term period, there are several approaches to the maintenance therapy, including: (1) maintenance of a triple drug combination at reduced doses compared to the early post-transplant period, (2) reduction or withdrawal of calcineurin inhibitor (CNI) or switch to rapamycin in patients sensitive to the nephrotoxic effect of CNIs, (3) early elimination of steroids, (4) reduction or discontinuation of anti-proliferative agents, (5) retaining full dose of calcineurin inhibitor or rapamycin associated to steroids in a double-drug regimen.

The overseeing of maintenance immunosuppressive therapy is often challenging, requiring careful attention to achieve a precise balance between two opposite conditions. Failure to maintain sufficient doses of immunosuppressive drugs can lead to the onset of acute rejection or chronic allograft nephropathy. Excessive immunosuppression increases the risk of drug-specific side effects and predisposes the patients to the development of infections and malignancy.

Pharmacokinetic monitoring of the immunosuppressive drugs is an important element in the management of the immunosuppressive treatment of transplanted patients because these drugs have a narrow therapeutic window and drug levels are unpredictable in patients because of inter- or intraindividual variation [1–3].

The 12-hour area under the time-concentration curve (AUC_{0-12}) most closely resembles a patient's drug exposure. Alternatively, the determination of the predose trough level (C_0), the concentration at $2\,h$ (C_2), and simplified models using either a three-point approach (C_1, C_2, and C_6) or a four-point approach (C_0, C_1, C_2, and C_4) are used in the attempt of realizing a good compromise between the pharmacokinetic parameters and clinically acceptable sampling times [4].

Several factors, including the nature of the transplanted organ, history of previous transplant, patient's age, time since transplantation, immunosuppressive protocol, drug-protein binding and the assay method used can influence the interpretation of the results.

Among the methods now available for the measurement of immunosuppressive drugs, high performance liquid chromatography is both sensitive and accurate and is still considered the reference assay for specific parent drug determinations. However, this assay requires a high level of technical skill, expensive equipment, relatively long turnaround times and extensive extraction procedures. Therefore, alternative automated techniques have been developed and used for routine clinical monitoring of immunosuppressive drugs (table 1) [5].

CNI

Original cyclosporine (CYA) (Sandimmun, Novartis) is a lipophilic polypeptide incorporated into olive oil (liquid formulation) or corn oil (gel cap)

Table 1. Automated immunoassay methods to analyze immunosuppressive drugs'
concentrations

Cyclosporine	
• FPIA:	fluorescence polarization immunoassay (TDX analyzer – Abbott)
• CYCLO-Trac:	whole-blood radioimmunoassay (RIA – DiaSorin)
• EMIT:	enzyme-multiplied immunoassay (Syva Company)
• CEDIA:	cloned enzyme donor immunoassay (Hoffman-La Roche)
Tacrolimus	
• ELISA:	enzyme-linked immunosorbent assay (DiaSorin)
• MEIA:	microparticle enzyme immunoassay (Abbott)
Mycophenolic acid	
• EMIT:	enzyme-multiplied immunoassay (Syva Company)
Sirolimus	
• MEIA:	enzyme immunoassay (IMX – Abbott)

with incomplete and variable bioavailability partly due to its dependence on bile
acids and emulsifiers for absorption. In the new formulation (Neoral) the drug
is incorporated into a microemulsion preconcentrate that is rapidly absorbed
without requiring the action of bile, enzymes or intestinal secretions [6].

There is a marked intra- and interpatient variability in CYA absorption
characteristics, although this variability is significantly lower with Neoral than
with the original Sandimmun formulation [7]. CYA absorption varies more
widely between and within patients during the first 4 h postdose [8].

Currently available data suggest that the maximal immunosuppressive
effect of CYA occurs during the first 4 h following Neoral administration [9].
C_0 correlates poorly with AUC_{0-4} in patients receiving Neoral, whereas C_2 is the
best timepoint predictor of AUC_{0-4} [10]. As full AUC_{0-4} is impractical in the
clinical setting, C_2 provides an effective single timepoint surrogate for AUC in
most patients allowing to identify whether patients are high, intermediate or low
absorbers of CYA and, consequently, to adjust the dosage accordingly.
Additional CYA concentration sampling beyond the C_2 timepoint such as C_4 or
C_6 may be required in a small proportion of patients who show markedly delayed
absorption of CYA and are termed slow absorbers. In the true slow absorber
both C_2 and C_6 would be low and Neoral dose should be increased according to
the C_2 value. In slow absorbers, C_6 is likely to be higher than C_2 and caution
should be exercised when increasing the dose of Neoral, to avoid toxicity.

C_0 monitoring of patients receiving Neoral distinguishes poorly between
those who will experience acute rejection and those who will remain rejection
free. On the other hand, higher C_2 levels are highly correlated with a lower risk
of acute rejection [11]. In one multicenter study [12] patients receiving Neoral

who had $C_2 > 1.5\,\mu g/ml$ at day 7 post-transplant experienced no rejection, whereas 58% of those with $C_2 < 1.5\,\mu g/ml$ experienced at least one rejection episode by day 14. Preliminary results from the international MO2ART study [13] have shown that individualizing Neoral dose based on C_2 monitoring achieves a biopsy-proven acute rejection rate $<10\%$ at 3 months.

Adjustment of Neoral dose based on C_2 monitoring does not appear to result in impaired early renal function in the short term. Preliminary results from the MO2ART study indicate that individualizing Neoral dose to target C_2 levels within the range $1.6–2.0\,\mu g/ml$ for the first month post-transplant does not adversely affect renal function in the first 3 months post-transplant. It is important to identify the slow absorbers to avoid inappropriate increases in Neoral dose in these patients, with consequent risk to renal function [13].

Patient management according to C_2 monitoring can improve graft renal function and can also reduce the incidence and severity of hypertension by identifying those who are receiving excessive CYA. Moreover, it has been reported that adult renal patients with chronic allograft nephropathy have significantly lower C_2 levels than those without graft impairment [14, 15].

The optimal long-term C_2 target to minimize the risk of chronic allograft nephropathy has not yet been established in a prospective study. Guideline Neoral C_2 target has been proposed for the first month post-transplant, with subsequent step-wise reductions in C_2 target levels over time. The optimal target for C_2 in adults with good initial graft function is $1.5–2.0\,\mu g/ml$. This target should be achieved by days 3–5. After the first month post-transplant, the C_2 target should be lowered progressively over time [16].

As to the accuracy of blood sampling, there is a window ranging 15 min before and after the 2 h timepoints during which the C_2 sample must be taken in order to remain within a 10% margin of error. Beyond this narrow period the level of error is considered too high and the information not useful for dose adjustment [17].

Tacrolimus (TAC) is a highly lipophilic compound with poor aqueous solubility. The oral absorption is incomplete and does not follow a regular pattern. The bioavailability of TAC administered orally ranges from as low as 4% to as high as 89% in different patients. The low bioavailability of the drug is attributed to poor distribution, poor gut permeability and back secretion by the P-glycoprotein pump and presystemic metabolism. Time to peak blood concentrations ranges from 0.5 to 6 h after oral administration [18]. Nevertheless, the intrapatient variability in the bioavailability of the drug, in systemic exposure, is considered to be low [19].

A good correlation was found between the blood concentration of TAC and clinical outcomes in transplanted patients. A significant relationship was described between both acute rejection, toxicity and TAC blood concentration

[20]. The relationship between low systemic exposure to TAC and acute rejection has been confirmed [21]. Furthermore, the relationship between TAC trough blood levels and adverse events has been investigated evaluating either the exposure to the drug over a period of time or the whole blood levels at the time of the event [22].

For the purpose of both efficacy and safety, exposure to TAC must be monitored during use. It has been shown that TAC AUC values on day 2 after transplantation correlate with clinical outcome: TAC AUC values were significantly lower in those patients who had experienced acute rejection than in those who had not [21, 23].

As routine use of AUC is clinically not practicable, several studies have shown a good correlation between AUC and trough concentration of TAC. Therefore, C_0 is used as a surrogate marker for systemic exposure (as reflected in the AUC). In addition, the whole-blood concentration-time profile of TAC is flat at 10–12 h after dosing. Thus, blood samples for monitoring taken within a ± 2-hour time window is considered to be equally predictive of exposure. Furthermore, evaluations of TAC whole-blood concentrations at other time-points (as 2 h postdosing) showed no better correlation with the AUC [24].

MMF

MMF is a morpholinoethylester prodrug of mycophenolic acid (MPA).

Following oral administration, MMF is rapidly and almost completely absorbed and then rapidly and completely converted via the glucuronyl transferase pathway to MPA, the active immunosuppressant drug. The sole metabolite of MPA is the glucuronide conjugate MPAG that is pharmacologically inactive in vitro.

Following administration of MMF, the plasma profile of MPA in fasting healthy subjects shows a rapid rise to achieve peak values at about one hour postdosing. Following the peak, the decline of plasma MPA concentration is initially rapid; subsequently, a secondary increase in plasma MPA concentration can be measured. This secondary peak is characteristically seen 6–12 h following oral administration of MMF, suggesting enterohepatic circulation of the drug substance.

Plasma MPA total AUC is the result of a number of factors, mainly hepatic metabolism and enterohepatic circulation. Because these processes might be differently affected under disease conditions, apparent systemic bioavailability may vary with the disease state [25].

MPA trough level and MPA AUC_{0-12} are significantly related to the incidence of biopsy-proven rejection, whereas the MMF dose is significantly related to the occurrence of some adverse events, in particular gastrointestinal symptoms [26].

MPA AUC is a better predictor of efficacy than predose trough levels because of the greater intra- and interpatient variability. The intrapatient coefficient of variation for both trough level and MPA AUC_{0-12} are greater in renal transplant patients with impaired early renal function, compared with patients with good early renal function [27].

Plasma protein binding of MPA is high, having a mean value in normal plasma of 98% and is independent of MPA concentration value. Only the free MPA concentrations have pharmacological activity. The MPA free fraction is one of the factors that appear to contribute significantly to interpatient variability [28]. A consequence of temporary increases in MPA-free fraction is a proportional increase in the oral clearance of the drug. Changes in the total concentration of MPA may not be associated with concurrent changes in free concentration. In transplant patients, several factors are known to cause decreased MPA binding to serum albumin. They include hypoalbuminemia, such as occurs in the early postsurgical period and uremia, either acute due to early delayed graft function or chronic due to allograft nephropathy [29]. Dose adjustments based on an increase in total concentration that occurs over time after transplantation, particularly in patients with early graft dysfunction, must be managed very carefully because interpretation of total concentration may be different in different clinical situations affecting protein binding.

Therapeutic drug monitoring of MMF is not generally accepted for the treatment of adult patients although there is an increasing evidence that this approach might help to reduce short-term and long-term side effects of MMF and long-term overimmunosuppression. MPA AUC_{0-12} has been shown to be the best predictor of MMF suppression of rejection in renal transplantation. On the basis of observations of both safety and efficacy, a dose of 1 g MMF b.i.d. represents the recommended starting dose in renal transplantation. Adjustment of dose should take account of MPA AUC_{0-12}. It appears that a total MPA AUC concentration in the ranges between 30 and 60 mg $\times$ h/l is a reasonable target for the early post-transplant period when MMF is prescribed with CYA. On the contrary, MPA trough level of less than 1 mg/l is associated with a higher incidence of rejection [30, 31].

As a full MPA AUC_{0-12} in clinical practice is not feasible routinely, abbreviated sampling schemes utilizing three to five blood samples have been proposed [32, 33].

In the clinical management of the transplanted patients, circumstances arise where the dose must be lowered to avoid drug toxicity such as bone marrow suppression or because of concurrent infection. Altering the dose of MMF within the first post-transplant year correlates with an increased incidence of acute rejection [34]. Moreover, the cumulative number of days with the MMF dose dropped below full dose is a significant predictor of acute rejection [35].

Rapamycin

Sirolimus (SRL) is a naturally occurring macrocyclic lactone. In the initial clinical trials SRL was administered orally as an oil-based solution. A solid formulation of SRL was more recently developed. Conversion of stable renal transplant recipients from the oil-based to the solid formulation results in similar AUC_{0-12} values. The tablet SRL formulation shows lower C_{max} values and a prolonged T_{max} indicating that the rate of absorption of the two formulations is not identical [36].

Following administration of multiple doses, in stable renal transplant patients, the half-life of SRL is about 62 h, the time to reach steady state is 6 days and a highly variable steady-state oral clearance has been calculated with an interpatient coefficient of variation of 50% and intrapatient coefficient of variation of 26%. The whole blood trough SRL concentrations are significantly correlated with AUC_{0-12} [37].

High performance liquid chromatography assay is currently the method of choice for determining SRL concentrations in patient's whole blood samples. An immunoassay (Imx) is currently in development.

SRL has been approved in United States and in the European Union for the prophylaxis of organ rejection in adult patients receiving renal transplant, at low-to-moderate immunological risk, with the recommendation that SRL be used initially in combination with CYA and steroids for the first 3 months. After this time, CYA should be progressively discontinued.

The usual dosage regimen of SRL is a 6-milligram oral loading dose, administered as soon as possible after transplantation, followed initially by a maintenance dose of 2 mg once daily. The SRL dose should then be individualized to obtain whole blood trough levels of 4–12 ng/ml. After withdrawal of CYA, a target whole blood trough range of 12–24 ng/ml is recommended. Optimally, adjustments in SRL dosage should be based on more than a single trough level obtained almost 5 days after a previous dosage change [38].

In pediatric patients and in subjects with hepatic impairment, the apparent clearance of SRL is significantly lower than in healthy volunteers and therefore the dosage of SRL should be reduced [39].

Other SRL-based immunosuppressive regimens with the association of TAC and MMF, respectively are under evaluation.

Everolimus (RAD) is an immunosuppressive macrolide bearing a stable 2-hydroxyethyl chain substitution at position 40 on the SRL (rapamycin) structure.

RAD, which has a greater polarity than SRL, was developed in an attempt to improve the pharmacokinetic characteristics of SRL, particularly to increase its oral bioavailability.

RAD and CYA show synergism in immunosuppression both in vitro and in vivo and therefore the drugs are intended to be given in combination after solid organ transplantation.

For the evaluation of the pharmacokinetics of RAD, specific high performance liquid chromatography assay and one enzyme-linked immunosorbent assay have been developed. Oral RAD is absorbed rapidly, reaching the C_{max} after 1.3–1.8 h and displaying a half-life of about 28 h, thus necessitating twice-daily dosing. Steady state is reached within 4 days. Steady-state peak and trough concentrations, and AUC_{0-12} are proportional to dosage [40]. In adults, RAD pharmacokinetic characteristics do not differ according to age, weight or sex, but body weight-adjusted dosages are necessary in children. Also RAD displays interindividual pharmacokinetic variability that can be explained by different activities of CYP3A4 and P-glycoprotein pump. In patients with hepatic impairment, the apparent clearance of RAD is significantly lower than in healthy volunteers, and therefore the dosage of RAD should be adequately reduced. Coadministration with CYA increases the exposure to RAD. The usual dosage regimen of RAD is 1.5–3 mg/day [41].

Because of the variable oral bioavailability and narrow therapeutic index of RAD, blood concentration monitoring seems to be important. The excellent correlation between steady-state trough concentration and AUC_{0-12} makes the former a simple and reliable index for monitoring RAD exposure. The target trough concentration of RAD should range between 3 and 15 µg/l in combination with CYA and steroids. A significantly increased risk of acute rejection was observed at RAD trough levels below 3 ng/ml. This is the lower therapeutic concentration limit when RAD is used with conventionally dosed CYA. RAD-related adverse events are manageable up to the trough levels of 15 ng/ml [42, 43].

Drug to Drug Interactions

Transplant recipients frequently receive a complex drug regimen including immunosuppressive agents and medications for complications or underlying pathology (hypertension, diabetes, dyslipidemia) with potential drug interactions. CNI and rapamycin are metabolized in the liver by cytochrome P450 3A4 (CYP3A4) isoenzyme [44]. Drugs or compounds that inhibit or induce the CYP3A4 isoenzyme may increase or decrease their blood levels, respectively, leading up to potentially toxic or, as opposite, subtherapeutic

Table 2. Drugs affecting cytochrome P450

Inducers	Inhibitors
Barbiturates	Clarithromycin
Carbamazepine	Clotrimazole
Cephalosporin	Danazol
Ciprofloxacin	Dexamethasone
Cyclosporine	Diltiazem
Glucocorticoids	Erythromycin
Isoniazid	Fluconazole
Octreotide	Itraconazole
Phenytoin	Ketoconazole
Rifabutin	Miconazole
Rifampin	Quinidine
Tacrolimus	Ranitidine
Ticlopidine	Verapamil
Troglitazone	

blood levels of immunosuppressants [1]. These interactions are summarized in table 2.

Both CYA and TAC are inhibitors of CYP3A4 isoenzyme. Corticosteroids are inducers of CYP3A4. SRL is a substrate for CYP3A4.

It was found that exposure to TAC increases by about 25% upon withdrawal of prednisolone. Mean serum creatinine levels also increased after steroid cessation and this can be due to the increase in TAC blood concentration [45].

When MMF is coadministered with TAC, the pharmacokinetics of TAC are unchanged, whereas the AUC values for MPA increases with time (from the first dose to month 3) for 1 and 2 g doses of MMF. This suggests that the dose of MMF might need to be reduced with time in order to maintain a stable exposure to MPA [46]. Higher MPA levels were observed when MMF was coadministered with TAC compared with the coadministration of CYA, at the same dose. The AUC_{0-12} values were also significantly higher [47]. When CYA is administered in combination with MPA in kidney transplant patients, a significant increase of MPA trough levels has been demonstrated after discontinuation of CYA, suggesting that coadministration of CYA reduces the exposure to MPA [48]. MPA concentrations in association with SRL are higher than those achieved when CYA is associated [49].

When SRL is used in combination with CYA for simultaneous administration, the mean C_{max} and AUC_{0-12} of SRL are significantly increased. This strong interaction is not evident when SRL is given 4 h after CYA administration. Because of the effect of CYA on SRL pharmacokinetics, it was recommended that SRL be taken 4 h after the administration of Neoral. Mean CYA C_{max} and AUC_{0-12} are not significantly affected when SRL is administered simultaneously with or 4 h after CYA. However, after the multiple-dose administration of SRL given 4 h after CYA in renal transplanted patients, over a period of 6 months, a reduced CYA oral dose clearance and the consequent need of lower doses of CYA to maintain target CYA blood concentration was observed [50]. When SRL is used in combination with TAC, an overall trend for reduced exposure to TAC with increasing doses of SRL has been documented, whereas TAC has no effect on the pharmacokinetics of SRL [51].

Drug Toxicity

All immunosuppressants have associated side effects that can be classified as therapeutic or toxic.

Aggressive therapy leading to excessive inhibition of T and B lymphocytes affects both innate and adaptive responses and can favor the emergence of infections or the development of neoplasms, in particular post-transplant lymphoproliferative disorders. This can be considered as a cumulative effect of the immunosuppression regimen. The toxic effect is more specific for a single drug.

The most important side effect associated with the use of CNI is nephrotoxicity. Acute changes include specific reduction of renal blood flow, renal plasma flow and glomerular filtration rate and tubular dysfunction. Chronic changes include progressive arteriolar damage, glomerular dropout, tubular atrophy and striped interstitial fibrosis. These problems can be preventable by appropriate dosage reduction of CNI early after transplant surgery. Hepatotoxicity also can be an issue with CNI use and therefore liver function should be monitored at regular intervals.

Among other nonrenal complications shared by CNI, CYA mainly contributes to hyperlipidemia, cardiovascular diseases, such as hypertension and cosmetic effects, whereas glucose intolerance and neurological symptoms occur more frequently in TAC-treated patients.

Gastrointestinal complications, in particular, abdominal pain, diarrhea, esophagitis, gastritis and gastrointestinal bleeding represent the most common adverse effects of MMF therapy. These problems may be overcome with a new enteric-coated formulation of MPA (ERL080A) which is in development. The drug has also been associated with bone marrow toxicity resulting in thrombocytopenia and leukopenia.

The side effect profile associated with SRL and RAD therapy includes thrombocytopenia, leukopenia and hyperlipidemia. The latter is the most serious effect of rapamycin and can exacerbate the CYA-induced hypercholesterolemia and the steroid-induced hypertriglyceridemia. Diarrhea, lymphocele, delayed wound healing and arthralgia represent less commonly observed toxic effects of this drug [52, 53].

Conclusions

The regulation of the immunosuppressive therapy after kidney transplantation is the most complex aspect of the management of transplanted patients. Every day the transplant clinician is challenged by the need to provide a sufficient immunosuppression to avoid or reduce the risk of rejection without exposing the patient to the risk of developing opportunistic infections or malignancy or toxic side effects.

The safety and efficacy profile of immunosuppressive therapy is limited within a narrow therapeutic window whose borders are represented by two clinical conditions such as rejection and drug-related toxicity. It is not possible to predict the effects of a given drug dose or blood concentration in a single patient and frequently the limits of the therapeutic window of a drug are very narrow and sometimes they even overlap.

The availability of several different drugs allows the clinicians to make multiple choices to individualize treatments according to the specific needs of a single patient.

The choice of one of the several immunosuppressive drugs available today requires careful evaluation of the preexisting disease and of the presence of any conditions that may affect the effects of the drug.

Pharmacokinetic monitoring of the immunosuppressive drugs is an important element in the management of these patients but cannot be considered as the unique driving factor and must be integrated with a careful surveillance and evaluation of all drug-related side effects.

References

1   Gaston R: Maintenance immunosuppression in the renal transplant recipient: An overview. Am J Kidney Dis 2001;38(suppl 6):S25–S35.
2   Filler G, Feber J, Lepage N, Weiler G, Mai I: Universal approach to pharmacokinetics monitoring of immunosuppressive agents in children. Pediatr Transplant 2002;6:411–418.
3   Shaw L, Holt D, Oellerich M, Meiser B, van Gelder T: Current issues in therapeutic drug monitoring of mycophenolate acid: Report of a round table discussion. Ther Drug Monit 2001;23: 305–315.
4   Gaspari F, Anedda MF, Signorini G, Remuzzi G, Perico N: Prediction of cyclosporine area under the curve using a three point sampling strategy after Neoral administration. J Am Soc Nephrol 1997;8:647–652.

5   Shaw L, Holt D, Keown P, Venkataramanan R, Yascoff R: Current opinions on therapeutic drug monitoring of immunosuppressive drugs. Clin Ther 1999;21:1632–1652.

6   Mueller EA, Kovarik JM, van Bree JB, Tetzloff W, Grevel J, Kutz K: Influence of a fat-rich meal on the pharmacokinetics of a new oral formulation of cyclosporine in a cross-over comparison with the market formulation. Pharm Res 1994;11:151–155.

7   Kahan BD, Dunn J, Fitts C, van Buren D, Wombold D, Pollack R, Carson R, Alexander JW, Choc M, Wong R: Reduced inter- and intrasubject variability in cyclosporine pharmacokinetics in renal transplant recipients treated with microemulsion formulation in conjunction with fasting, low-fat meals, or high-fat meals. Transplantation 1995;59:505–511.

8   Johnston A, David OJ, Cooney GF: Pharmacokinetics validation of Neoral absorption profiling. Transplant Proc 2000;32:S53–S56.

9   Steiner CM, Murray JJ, Wood AJ: Inhibition of stimulated interleukin-2 production in whole blood: A practical measure of cyclosporine effect. Clin Chem 1999;45:1477–1484.

10   Mahalati K, Belitsky P, Sketris I, West K, Panek R: Neoral monitoring by simplified sparse sampling area under the concentration–time curve. Transplantation 1999;68:55–62.

11   Pescovitz MD, Barbeito R, Simulect US01 Study Group: Two-hour post-dose cyclosporine level is a better predictor than trough level of acute rejection of renal allografts. Clin Transplant 2002;16:378–382.

12   Canadian Neoral Renal Transplantation Study Group: Absorption profiling of cyclosporine microemulsion (Neoral) during the first 2 weeks after renal transplantation. Transplantation 2001;72:1024–1032.

13   Pfeffer P, Stefoni S, Agost Carreno C, Thervet E, Fornarion S, Keown P, MO2ART Study Group: Monitoring of 2-hour Neoral absorption In renal transplantation (MO2ART): Interim analysis shows low incidence of acute rejection in the early post-graft period. Abstract 1037 Am Soc Transpl Congress, Washington DC, Apr 26–May 1, 2002.

14   Citterio F, Scata MC, Borzi MT, Pozzetto U, Castagneto M: C2 single-point sampling to evaluate cyclosporine exposure in long-term renal transplant recipients. Transplant Proc 2001;33: 3133–3136.

15   Nashan B, Cole E, Levy G, Thervet E: Clinical validation studies of Neoral C2 monitoring: A review. Transplantation 2002;73(suppl):S3–S11.

16   Levy G, Thervet E, Lake J, Uchida K, on behalf of the CONCERT Group: Patient management by Neoral C2 monitoring: An international consensus statement. Transplantation 2002;73(suppl 9): S12–S18.

17   Kahan BD, Keown P, Levy GA, Johnston A: Therapeutic drug monitoring of immunosuppressant drugs in clinical practice. Clin Ther 2000;24:1–21.

18   Venkataramanan R, Swaminathan A, Prasad T, Jain A, Zuckerman S, Warty V, McMichael J, Lever J, Burckart G, Starzl T: Clinical pharmacokinetics of Tacrolimus. Clin Pharmacokinet 1995;29: 404–430.

19   Undre NA: Pharmacokinetics of Tacrolimus-based therapies. Nephrol Dial Transplant 2003; 18(suppl 1):i12–i15.

20   Kershner RP, Fitzsimmons WE: Relationship of FK506 whole blood concentrations and efficacy and toxicity after liver and kidney transplantation. Transplantation 1996;62:920–926.

21   Undre NA, van Hooff J, Christiaans M, Vanrenterghem Y, Donck J, Heemann U, Kohnle M, Zanker B, Land W, Morales JM, Andres A, Schafer A, Stevenson P: Low systemic exposure to tacrolimus correlates with acute rejection. Transplant Proc 1999;31:296–298.

22   Mayer AD, Dmitrewski J, Squifflet JP, Besse T, Grabensee B, Klein B, Eigler FW, Heemann U, Pichlmayr R, Behrend M, Vanrenterghem Y, Donck J, van Hooff J, Christiaans M, Morales JM, Andres A, Johnson RW, Short C, Buchholz B, Rehmert N, Land W, Schleibner S, Forsythe JL, Talbot D, Pohanka E: Multicenter randomized trial comparing tacrolimus (FK506) and cyclosporine in the prevention of renal allograft rejection: A report of the European Tacrolimus Multicenter Renal Study Group. Transplantation 1997;64:436–443.

23   Undre NA, Stevenson P, Schafer A: Pharmacokinetics of Tacrolimus: Clinically relevant aspects. Transplant Proc 1999;31(suppl 7):S21–S24.

24   Bottiger Y, Undre NA, Sawe J, Stevenson PJ, Ericzon BG: Effect of bile flow on the absorption of Tacrolimus in liver allograft transplantation. Transplant Proc 2002;34:1544–1545.

25 Bullingham RES, Nicholls A, Hale M: Pharmacokinetics of mycophenolate mofetil (RS61443): A short review. Transplant Proc 1996;28:925–929.

26 van Gelder T, Hilbrands LB, Vanrenterghem Y, Weimar W, de Fijter JW, Squifflet JP, Hené RJ, Verpooten GA, Navarro MT, Hale MD, Nicholls AJ: A randomized double-blind multicenter plasma concentration controlled study of the safety and efficacy of oral mycophenolate mofetil for the prevention of acute rejection after kidney transplantation. Transplantation 1998;68: 261–266.

27 Shaw LM, Kaplan B, DeNofrio D, Korecka M, Brayman KL: Pharmacokinetics and concentration-control investigations of mycophenolic acid in adults after transplantation. Ther Drug Monit 2000;22:14–19.

28 Nowak I, Shaw LM: Mycophenolic acid binding to human serum albumin: Characterization and relation to pharmacodynamics. Clin Chem 1995;41:1011–1017.

29 Shaw LM, Korecka M, Aradhye S, Grossman R, Bayer L, Innes C, Cucciara A, Barker C, Naji A, Nicholls A, Brayman K: Mycophenolic acid area under the curve values in African American and Caucasian renal transplant patients are comparable. J Clin Pharmacol 2000;40:624–633.

30 Nicholls AJ: Opportunities for therapeutic monitoring of mycophenolate mofetil dose in renal transplantation suggested by the pharmacokinetic/pharmacodynamic relationship for mycophenolic acid and suppression of rejection. Clin Biochem 1998;31:329–333.

31 Shaw LM, Nicholls A, Hale M, Armstrong VW, Oellerich M, Yatscoff R, Morris RE, Holt DW, Venkataraman R, Haley J, Halloran P, Ettenger R, Keown P, Morris RG: Therapeutic monitoring of mycophenolic acid: A consensus panel report. Clin Biochem 1998;31:317–322.

32 Willis C, Taylor PJ, Salm P, Tett SE, Pillans PI: Evaluation of limited sampling strategies for estimation of 12-hour mycophenolic acid area under the plasma concentration-time curve in adult renal transplant patients. Ther Drug Monit 2000;22:549–554.

33 Filler G: Abbreviated mycophenolic acid AUC from C0, C1, C2, and C4 is prefeerable in children after renal transplantation on mycophenolate mofetil and tacrolimus therapy. Transpl Int 2004;17:120–125.

34 Pelletier RP, Akin B, Henry ML, Bumgardner GL, Elkhammas EA, Rajab A, Ferguson RM: The impact of mycophenolate mofetil dosing patterns on clinical outcome after renal transplantation. Clin Transplant 2003;17:200–205.

35 Knoll GA, Mac Donald I, Khan A, van Walraven C: Mycophenolate mofetil dose reduction and the risk of acute rejection after renal transplantation. J Am Soc Nephrol 2003;14: 2381–2386.

36 Kelly PA, Napoli KL, Kahan BD: Conversion from liquid to solid rapamycin formulations in stable renal allograft transplant recipient. Biopharm Drug Dispos 2000;20:249–253.

37 Zimmerman JJ, Kahan BD: Pharmacokinetics of sirolimus in stable renal transplant patients after multiple oral dose administration. J Clin Pharmacol 1997;37:405–415.

38 Holt DW, Denny K, Lee TD, Johnston A: Therapeutic monitoring of sirolimus: Its contribution to optimal prescription. Transplant Proc 2003;35(suppl 3):S157–S161.

39 Oellerich M, Armstrong V, Streit F, Weber L, Tonshoff B: Immunosuppressive drug monitoring of sirolimus and cyclosporine in pediatric patients. Clin Biochem 2004;37:624–628.

40 Kahan BD, Wong RL, Carter C, Katz SH, von Fellenberg J, van Buren CT, Appel-Dingemanse S: A phase I study of a 4-week course of SDZ-RAD (RAD) quiescent cyclosporine-prednisone-treated renal transplant recipients. Transplantation 1999;68:1100–1106.

41 Kovaric JM, Kalbag J, Figueiredo J, Rouilly M, Frazier OL, Rordorf C: Differential influence of two cyclosporine formulations on everolimus pharmacokinetics: A clinically relevant pharmacokinetic interaction. J Clin Pharmacol 2002;42:95–99.

42 Kirchner GI, Meier-Wiedenbach I, Manns MP: Clinical pharmacokinetics of everolimus. Clin Pharmacokinet 2004;43:83–95.

43 Kovarik JM, Kaplan B, Tedesco Silva H, Kahan BD, Dantal J, Vitko S, Boger R, Rordorf C: Exposure-response relationship for everolimus in de novo kidney transplantation: Defining a therapeutic range. Transplantation 2002;73:920–925.

44 Karanam BV, Vincent SH, Newton DJ, Wang RW, Chiu SH: FK506 metabolism in human liver microsomes: Investigation of the involvment of cytochrome P450 isoenzymes other than CYP3A4. Drug Metab Dispos 1994;22:811–814.

45   van Duijnhoven EM, Boots JMM, Christiaans M, Undre NA, van Hooff J: Tacrolimus trough levels increase after steroid withdrawal. 2nd International Congress on Immunosuppression, San Diego, 2001, 133 (abstract P88).

46   Undre NA, van Hooff J, Christiaans M, Vanrenterghem Y, Donck J, Heeman U, Kohnle M, Zanker B, Land W, Morales JM, Andres A, Schafer A, Stevenson P: Pharmacokinetics of FK506 and mycophenolic acid after the administration of a FK506-based regimen in combination with mycophenolate mofetil in kidney transplantation. Transplant Proc 1998;30:1299–1302.

47   Zucker K, Rosen A, Tsaroucha A, de Faria L, Roth D, Ciancio G, Esquenazi V, Burke G, Tzakis A, Miller J: Augmentation of mycophenolate mofetil pharmacokinetics in renal transplant patients receiving Prograf and Cellcept in combination therapy. Transplant Proc 1997;29:334–336.

48   Gregoor PJ, de Sevaux RG, Hene RJ, Hesse CJ, Hilbrands LB, Vos P, van Gelder T, Hoitsma AJ, Weimar W: Effect of cyclosporine on mycophenolic acid trough levels in kidney transplant recipient. Transplantation 1999;68:1603–1606.

49   Kreis H, Cisterne JM, Land L, Wrammer L, Squifflet JP, Abramowicz D, Campistol JM, Morales JM, Grinyo GM, Mourad G, Berthoux FC, Brattstrom C, Lebranchu Y, Vialtel P: Sirolimus in association with mycophenolate mofetil in renal transplant recipients. Transplantation 2000;69:1252–1260.

50   MacDonald A, Scarola J, Burke JT, Zimmerman JJ: Clinical pharmacokinetics and therapeutic drug monitoring of sirolimus. Clin Ther 2000;22(suppl B):B101–B121.

51   Undre Na: Pharmacokinetics of tacrolimus-based combination therapies. Nephrol Dial transplant 2003;18(suppl 1):i12–i15.

52   Shaw LM, Kaplan B, Kaufman D: Toxic effects of immunosuppressive drugs: Mechanism and strategies for controlling them. Clin Chem 1996;42:1316–1321.

53   Kahan B, Ponticelli C: Established immunosuppressive drugs: Clinical and toxic effects; in Kahan B, Ponticelli C (eds): Principles and Practice of Renal Transplantation. London, Martin Dunitz, 2000, pp 349–414.

Stefano Chiaramonte, MD
Department of Nephrology, Dialysis and Transplantation
St. Bortolo Hospital, Viale Rodolfi, 37, IT–36100 Vicenza (Italy)
Tel. +39 0444 993650, Fax +39 0444 993973, E-Mail stefano.chiaramonte@ulssvicenza.it

Ronco C, Chiaramonte S, Remuzzi G (eds): Kidney Transplantation: Strategies to Prevent
Organ Rejection. Contrib Nephrol. Basel, Karger, 2005, vol 146, pp 87–94

Chronic Allograft Nephropathy

A Multiple Approach to Target Nonimmunological Factors

Piero Ruggenenti

Department of Medicine and Transplantation, Ospedali Riuniti, Azienda Ospedaliera,
and Mario Negri Institute for Pharmacological Research, Bergamo, Italy

Introduction

In the past two decades, the short-term results of renal transplantation have
dramatically improved [1]. Better short-term outcomes, however, have not been
paralleled by an acceptable improvement in the long-term allograft survival [2].
When death with a functioning graft is excluded as a cause of late renal
allograft loss, most grafts fail after a period of renal function deterioration [3],
which has been attributed to a process of progressive renal structural injury
called 'chronic transplant nephropathy' [4]. Although the functional and
morphological findings are well characterized, the pathophysiological mechanisms leading to graft deterioration secondary to the evolution of this process
are poorly understood. Potential risk factors, beside immune events, include
hypertension, inadequate functional nephron mass, drug toxicity, de novo or
recurrent progressive renal disease [5].

Angiotensin II, the principal effector of the renin-angiotensin cascade,
stimulates many physiological responses that support blood pressure and renal
function [6]. Moreover, in addition to being a powerful vasoconstrictor,
angiotensin II is a potent mediator of cellular proliferation and extracellular
matrix protein synthesis and accumulation [6]. These effects contribute to
progressive fibrotic diseases in various organ systems. Thus, abnormal generation of angiotensin II has been implicated in the pathogenesis of hypertension,
cardiovascular diseases, and progressive renal diseases [6, 7]. Recently, experimental evidence of similar pathogenic mechanisms of progression has been
provided in chronic transplant nephropathy and other chronic renal diseases, in
which angiotensin II is recognized as a central effector [8].

Pharmacological interruption of the renin-angiotensin system (RAS) with angiotensin-converting enzyme (ACE) inhibitors is increasingly advocated as a standard therapeutic intervention for patients with chronic renal disease, regardless of whether systemic hypertension is an associated feature [7, 9]. The advent of orally active angiotensin receptor blockers has increased therapeutic options for inhibiting the RAS in patients with progressive renal disease [7]. Nevertheless, despite ample evidence to support the recommendation of RAS blockade therapy as the standard of care for strategies aimed at preserving renal function in chronic renal disease and the well-established antihypertensive effect of these drugs, the use of RAS blockers in renal transplantation has been quite limited. Indeed for years nephrologists have been reluctant to use them in renal transplant recipients because of a number of alarming reports on their possibility to induce renal insufficiency [10, 11].

Reduced Nephron Mass as a Cause of Chronic Allograft Nephropathy

The two most common causes of late graft loss after the first year of renal transplantation are chronic allograft nephropathy – a clinicopathological entity whose exact nature is not yet defined and patient death [3].

Transplanted patients with chronic graft nephropathy exhibit a gradual and progressive deterioration of renal function in association with proteinuria and arterial hypertension [3]. Histological examination of kidneys with chronic allograft nephropathy shows arterial intimal and medial fibrosis and arteriolar insudative lesions, glomerulosclerosis, and interstitial fibrosis with tubular atrophy [12]. These functional and structural changes of chronic renal allograft failure show similarities with those observed in other forms of chronic progressive renal disease in which inadequate functioning of the nephron mass has been considered the key event [13]. Indeed, transplantation of a single kidney theoretically supplies half the number of nephrons commonly available to a healthy subject. This implies an increased workload per nephron to maintain body homeostasis [14]. The transplanted kidney is also subject to further reduction in the pool of functioning nephrons due to ischemic injury, acute rejection and chronic cyclosporine toxicity [13]. In experimental animals, a too small number of nephrons due to a small renal mass for acquired or innate reasons, triggers a self-perpetuating cycle of events, the hallmark of which is excessive urinary protein excretion followed by interstitial and glomerular inflammation and scarring [9]. Hemodynamic determinants of subsequent renal injury in this setting are enhanced intra-glomerular pressure and flow, closely involved in the development of renal structural damage [15]. Glomerular hypertension enhances filtration of macromolecules across the capillary barrier, which are then largely reabsorbed by proximal tubuli [9, 16]. This tubular cell activation up-regulates genes for inflammatory and vasoactive proteins that, in the long run, contribute additionally to renal scarring [9]. All these mechanisms could logically operate in a single kidney graft.

Slowing Progression of Chronic Allograft Nephropathy: The Role of RAS Blockade

Blockade of the RAS with ACE inhibitors or Angiotensin II (A II) receptor antagonists reduces urinary protein excretion and protects against renal structural injury better than conventional therapy in nontransplant rat models of chronic renal disease due to lower than normal nephron numbers [7], and in humans with proteinuric renal diseases [9]. Collectively, trials in diabetic and nondiabetic progressive renal disease showed that, at comparable levels of blood pressure control, ACE inhibitors slowed the rate at which renal function is lost better than other anti-hypertensive agents [9, 17]. This reno-protective effect was consistently associated with a substantial limitation of urinary protein excretion. The possibility of a common pathogenic mechanism for progressive renal diseases suggested that therapies already found effective in slowing the progression of several immune and nonimmune renal diseases unrelated to transplantation could also limit chronic renal allograft dysfunction. Experimental studies have shown that A II receptor blockade significantly helped to prevent chronic graft injury in the Fisher-Lewis rat model of late renal allograft failure [18–21]. Alleviation of glomerular capillary hypertension with A II receptor antagonist treatment was associated with a reduction in proteinuria and the prevention of supervening chronic glomerular and tubulointerstitial injury [18, 19, 21]. With this promising background, the use of RAS blockers to protect renal function in kidney transplant patients was a logical extension.

However, there has been some concern in the transplant community that these agents may impair graft function by potentiating the reduction of glomerular filtration rate (GFR) already caused by cyclosporine A and precipitating acute renal failure if administered to renal transplant recipients of a single kidney [10, 11]. Thus so far, no appropriate controlled studies aimed at evaluating the antiproteinuric and renoprotective effect of RAS blockers are available. In 10 patients with post-transplant hypertension, on immunosuppressive therapy with azathioprine and prednisone, who discontinued their previous anti-hypertensive medications 6–72 months after surgery, fosinopril taken for 12 months normalized blood pressure, and progressively reduced the 24-hour urinary protein excretion, but also decreased GFR values which returned to baseline when the drug was discontinued [22]. In 22 patients with transplant nephrotic syndrome on double or triple immunosuppressive therapy, incremental doses of enalapril for one year resulted in a significant fall in mean daily proteinuria without changes in renal function, measured as predicted creatinine clearance by the Cockroft-Gault equation [23]. The short-term efficacy and safety of ACE inhibition in kidney transplant recipients was confirmed in 8 patients in whom microalbuminuria improved with nonsignificant change in mean blood pressure or GFR after a 3-month therapy [24]. A significant decline in urinary protein excretion was also reported in 76% of proteinuric transplant patients given enalapril or captopril for an average of 21 months [25]. The favorable effect of ACE inhibition on proteinuria was associated with stabilization of renal function in 62% of cases. More recently, a clinicopathological study of post-transplant IgA nephropathy showed a beneficial effect of the ACE inhibitor trandolapril, added to their current anti-hypertensive regimen with nifedipine or amlodipine, in lowering urinary protein excretion during 12–16 month follow-up, without any graft function deterioration [26].

The above studies, however, were not designed to compare the anti-proteinuric effect of ACE inhibitors and other anti-hypertensive agents. In a prospective randomized study in renal transplant recipients, lisinopril alone or in association with furosemide, or nifedipine alone or combined with atenolol did not modify urinary albumin excretion nor GFR during 2.5 years of follow-up, despite similar effect on lowering blood pressure [27]. In 13 transplanted patients with proteinuria exceeding 0.5 g/dl, neither an ACE inhibitor nor CCB treatment had any significant effect on urinary protein excretion [28]. Moreover, a significant reduction of proteinuria was found at the end of a 2-month treatment with perindopril in transplant patients with stable renal function as compared to nifedipine [29].

Together these studies led to inconsistent results as to the effectiveness of ACE inhibitors to reduce protein excretion in renal transplant recipients. It should, however, be pointed out that most of these studies were primarily aimed just to assess the lowering blood pressure efficacy of ACE inhibitors and not specifically the anti-proteinuric effect of these drugs.

Recent findings with A II receptor blockers have been more promising. Indeed, after initial anedoctal observations that A II blockade with losartan normalized urinary protein excretion in renal transplant patients with severe proteinuria [30], the potential effect of this new class of drugs in clinical transplantation have been explored more systematically. Nevertheless, appropriate controlled studies are scanty. As a part of a trial investigating primarily the efficacy and safety of losartan in the treatment of hypertension in renal transplant recipients, the effect on urinary proteins excretion was also examined [31]. In the 67 patients who completed the 12-week study period, proteinuria significantly decreased as early as after 4 weeks of therapy with the A II antagonist, with a further progressive decline thereafter. This was associated with a slight, nonclinically significant increase in serum creatinine concentration at 4 weeks, then remaining stable until study completion. Treatment with losartan also reduced proteinuria in 14 transplant patients with chronic allograft nephropathy during 8 weeks of follow-up, without any negative effect on graft function [32]. The anti-proteinuric effect of losartan was also confirmed in few patients that in a retrospective chart analyses of 642 renal transplant recipients had proteinuria greater than 2 g/day at least one month post-transplant and were given the A II blocker for at least 6 months [33]. More recently, in 11 renal transplant recipients with overt proteinuria either a reduction or a stabilization in urinary protein excretion was detected after losartan administration for a mean period of 14 months at the dose of 25–100 mg/day according to the anti-hypertensive response obtained [34]. Moreover, new onset of proteinuria was not documented during the treatment period in any of the other 7 nonproteinuric patients [34]. No correlation was found between the reduction in proteinuria and the decrease in mean arterial pressure, suggesting that the anti-proteinuric effect of the A II blocker was independent of blood pressure changes. Overall graft function remained stable. Others have speculated that A II antagonists can affect chronic allograft nephropathy, besides reducing proteinuria, by modulating the generation of transforming growth factor B-1 (TGF-B1), a key fibrogenic cytokine involved in the fibrosis of a number of chronic diseases of the kidney and other organs [35]. This possibility rests on the observation that in transplant recipients with chronic allograft nephropathy losartan decreased the plasma levels of TGF-B1 to a value comparable to that in transplant patients with nonclinical evidence of chronic nephropathy [32]. Since reduction of the plasma concentration of the cytokine by A II antagonism occurred in proteinuric and nonproteinuric patients, a direct effect of losartan treatment on TGF-B production has been hypothesized. These results confirmed recent data from experimental animals and in vitro studies in which ACE inhibitory and A II antagonism decreased the synthesis and secretion of renal TGF-B1 and prevented the development of interstitial fibrosis and subsequent chronic nephropathy [36]. Up to now the limited number of studies with A II antagonists on the progression of chronic allograft nephropathy have mainly focused on the anti-proteinuric effect of these drugs, usually in a short period of follow-up. Unfortunately, as yet we have no information about the impact of A II antagonism on progressive renal graft dysfunction and ultimately on long-term graft survival.

Conclusion

The potential renoprotective effect of RAS blockers in transplantation rests on their documented efficacy in diabetic and nondiabetic proteinuric nephropathies in which the reduced number of functioning nephrons is the key event in progressive renal injury, as it has been postulated for single kidney transplant. Available so far are only short-term follow-up data with this class of compounds in transplant recipients, showing their ability to lower urinary protein excretion, which, however, are not enough for predicting the impact of these drugs on kidney graft survival in the long term. This issue merits investigation in clinical trials designed ad hoc. Nevertheless, our current knowledge converges to suggest that blockade of RAS should be regarded as a first line of therapy even in transplant recipients. These agents may provide a valuable alternative means of simultaneously addressing different problems of transplant patients that until recently have been solved with a multidrug approach or in the case of post-transplant erythrocytosis by nonpharmacological approaches. Moreover, based on studies in the general population showing that proteinuria is an important risk factor for cardiovascular mortality [37] and that RAS blockers reduce proteinuria and cardiovascular mortality, this class of drugs may also be useful for reducing or preventing a major cause of morbidity and death among renal transplant recipients [38]. Thus, it is time now for nephrologists to stably move RAS blockers into the renal transplant area. Appropriate monitoring of serum creatinine, potassium and hemoglobin would help in optimizing the use of these drugs in transplant patients. Studies are also needed to assess whether other agents with anti-proteinuric and cardioprotective properties such as 3-hydroxy-3-methyl-glutamyl coenzyme A, inhibitors may be safely used in combination with ACE inhibitors and/or A II antagonists to further improve long-term patient and graft survival.

References

1   Terasaki P, Yuge J, Cecka JM, Gjertson DW, et al: Thirty-year trends in clinical kidney transplantation; in Terasaki P, Cecka JM (eds): Clinical Transplants 1993. Los Angeles, UCLA Tissue Typing Laboratory, 1994, pp 553–562.
2   Hariharan S, Johnson CP, Bresnahan BA, Taranto SE, et al: Improved graft survival after renal transplantation in the United States, 1988 to 1996. N Engl J Med 2000;342:605–612.
3   Paul L: Chronic renal transplant loss. Kidney Int 1995;47:1491–1499.
4   Hayry P, Isoniemi H, Ylmaz S, Mennander A, et al: Chronic allograft rejection. Immunol Rev 1993;134:33–79.
5   Paul LC, Benediktsson H: Chronic transplant rejection: Magnitude of the problem and pathogenic mechanisms. Transplant Rev 1993;134:5–19.
6   Perico N, Remuzzi G: Angiotensin II receptor antagonists and treatment of hypertension and renal disease. Curr Opin Nephrol Hypertens 1998;7:571–578.

7   Taal MW, Brenner BM: Renoprotective benefits of RAS inhibition: From ACEI to angiotensin II antagonists. Kidney Int 2000;57:1803–1817.

8   Remuzzi G, Perico N: Protecting single-kidney allografts from long-term functional deterioration. J Am Soc Nephrol 1998;9:1321–1332.

9   Remuzzi G, Bertani T: Pathophysiology of progressive nephropathies. N Engl J Med 1998; 339:1448–1456.

10   Curtis JJ, Luke RG, Whelchel JD, Diethelm AG, et al: Inhibition of angiotensin-converting enzyme in renal transplant recipients with hypertension. N Engl J Med 1983;308:377–381.

11   Ahmad T, Coulthard MG, Eastham EJ: Reversible renal failure due to the use of captopril in a renal allograft recipient treated with cyclosporin. Nephrol Dial Transplant 1989;4:311–312.

12   Solez K, Axelsen RA, Benediktsson H, Burdick JF, et al: International standardization of criteria for the histologic diagnosis of renal allograft rejection: The Banff working classification of kidney transplant pathology. Kidney Int 1993;44:411–422.

13   Mackenzie HS, Brennner BM: Antigen-independent determinants of late renal allograft outcome: The role of renal mass. Curr Opin Nephrol Hypertens 1996;5:289–296.

14   Brenner BM, Cohen RA, Milford EL: In renal transplantation, one size may not fit all. J Am Soc Nephrol 1992;3:162–169.

15   Brenner BM: Hemodynamically mediated glomerular injury and the progressive nature of glomerular disease. Kidney Int 1983;23:647–655.

16   Remuzzi G: Abnormal protein traffic through the glomerular barrier induces proximal tubular cell dysfunction and causes renal injury. Curr Opin Nephrol Hypertens 1995;4:339–342.

17   Ruggenenti P, Remuzzi G: Angiotensin-converting enzyme inhibitor therapy for non-diabetic progressive renal disease. Curr Opin Nephrol Hypertens 1997;6:489–495.

18   Mackenzie HS, Ziai F, Nagano H, Azuma H, et al: Candesartan cilexetil reduces chronic renal allograft injury in Fisher – Lewis rats. J Hypertens 1997;15(suppl 4):S21–S25.

19   Benediktsson H, Chea R, Davidoff A, Paul LC: Antihypertensive drug treatment in chronic renal allograft rejection in the rat. Effect on structure and function. Transplantation 1996;62: 1634–1642.

20   Amuchastegui SC, Azzollini N, Mister M, Pezzotta A, et al: Chronic allograft nephropathy in the rat is improved by angiotensin II receptor blockade but not by calcium channel antagonism. J Am Soc Nephrol 1998;9:1948–1955.

21   Ziai F, Nagano H, Kusaka M, Coito AJ, et al: Renal allograft protection with losartan in Fisher – Lewis rats: Hemodynamics, macrophages, and cytokines. Kidney Int 2000;57:2618–2625.

22   Bochicchio T, Sandoval G, Ron O, Perez-Grovas H, et al: Fosinopril prevents hyperfiltration and decreases proteinuria in post-transplant hypertensives. Kidney Int 1990;38:873–879.

23   Rell K, Linde J, Morzycka-Michalik M, Gaciong Z, et al: Effect of enalapril on proteinuria after kidney transplantation. Transpl Int 1993;6:213–217.

24   Mulhern J, Lipkowitz GS, Braden GL, Madden RL, et al: Association of post-renal transplant erythrocytosis and microalbuminuria: Response to angiotensin-converting enzyme inhibition. Am J Nephrol 1995;15:318–322.

25   Oppenheimer F, Flores R, Cofan F, Campistol JM, et al: Treatment with angiotensin-converting enzyme inhibitors in renal transplantation with proteinuria. Transplant Proc 1995;27: 2235–2236.

26   Oka K, Imai E, Moriyama T, Akagi Y, et al: A clinicopathological study of IgA nephropathy in renal transplant recipients: Beneficial effect of angiotensin-converting enzyme inhibitor. Nephrol Dial Transplant 2000;15:689–695.

27   Mourad G, Ribstein J, Mimran A: Converting-enzyme inhibitor versus calcium antagonist in cyclosporin-treated renal transplants. Kidney Int 1993;43:419–425.

28   Van der Schaaf MR, Hene RJ, Floor M, Blankestijn PJ, et al: Hypertension after renal transplantation: Calcium channel or converting enzyme blockade? Hypertension 1995;25:77–81.

29   Grekas D, Dioudis C, Kalevrosoglou I, Alivanis P, et al: Renal hemodynamics in hypertensive renal allograft recipients: Effects of calcium antagonists and ACE inhibitors. Kidney Int 1996;49 (suppl 55):S97–S100.

30   Navarro JF, MAcia ML, Garcia J: Control of severe proteinuria with losartan after renal transplantation. Am J Nephrol 1998;18:261–262.

31   Del Castillo D, Campistol JM, Guirado L, Capdevilla L, et al: Efficacy and safety of losartan in the treatment of hypertension in renal transplant recipients. Kidney Int 1998;54(suppl 68): S135–S139.

32   Campistol JM, Inigo P, Jimenez W, Lario S, et al: Losartan decreases plasma levels of TGF-b1 in transplant patients with chronic allograft nephropathy. Kidney Int 1999;56:714–719.

33   Stigant CE, Cohen J, Vivera M, Zaltzman JS: ACE inhibitors and angiotensin II antagonists in renal transplantation: An analysis of safety and efficacy. Am J Kidney Dis 2000;35:58–63.

34   Calvino J, Lens XM, Romero R, Sanchez-Guisande D: Long-term anti proteinuric effect of Losartan in renal transplant recipients treated for hypertension. Nephrol Dial Transplant 2000;15:82–86.

35   Border WA, Noble NA: Transforming growth factor-b in tissue fibrosis. N Engl J Med 1994; 51:646–671.

36   Border WA, Noble NA: Interactions of transforming growth factor-b and angiotensin II in renal fibrosis. Hypertension 1998;31:181–188.

37   Kannel WB, Stampfer MJ, Castelli WP, Verter J: The prognostic significance of proteinuria: The Framingham study. Am Heart J 1984;108:1347–1352.

38   Ojo AO, Hanson JA, Wolfe RA, Leichtman AB, et al: Long-term survival in renal transplant recipients with graft function. Kidney Int 2000;57:307–313.

Piero Ruggenenti, MD
Mario Negri Institute for Pharmacological Research
Via Gavazzeni 11, IT–24125 Bergamo (Italy)
Tel. +39 035 319888, Fax +39 035 319331, E-Mail manuelap@marionegri.it

Ronco C, Chiaramonte S, Remuzzi G (eds): Kidney Transplantation: Strategies to Prevent
Organ Rejection. Contrib Nephrol. Basel, Karger, 2005, vol 146, pp 95–104

··············

Transplantation Tolerance

A Complex Scenario Awaiting Clinical Applicability

Mohamed H. Sayegh[a], *Norberto Perico*[b,c], *Giuseppe Remuzzi*[b]

[a]Transplantation Research Center, Brigham and Women's Hospital and Children's
Hospital Boston, Harvard Medical School, Boston, Mass., USA; [b]Department of
Medicine and Transplantation, Ospedali Riuniti di Bergamo, Mario Negri Institute
for Pharmacological Research, Bergamo, Italy, and [c]Center for Research on Organ
Transplant 'Chiara Cucchi De Alessandri & Gilberto Crespi', Ranica (Bergamo), Italy

Abstract

Organ transplantation is now firmly established as the therapy of choice for end-stage
organ failure. Specific immunological tolerance of transplant recipients towards their foreign
organ or tissue grafts is a goal that has been sought by transplant biologists for almost
50 years following the original description of the phenomenon in experimental animals by
Medawar and colleagues. Since that time, a wealth of experimental data has accumulated
relating to strategies for extending allograft survival and function. Recent studies have shed
new light on the molecular and cellular basis of transplant rejection and have better defined
the mechanisms of allograft tolerance with particular emphasis on a role for regulatory
T cells. Still, the question remains of how near we are to the day when long-term tolerance
of engrafted organs or tissues will be a clinical reality. Recently, clinical trials to explore
pilot tolerance protocols in humans have been initiated under the auspices of the Immune
Tolerance Network (www.immunetolerance.org). In this review we will highlight the
promise and challenges of making transplantation tolerance a clinical reality.

Over the last three decades, transplantation has become the preferred
approach for the treatment of failure of the heart, liver, kidneys and lungs.
Recently, a progressive improvement of allograft survival, in particular kidney
allografts has been reported [1]. Intriguingly, this improvement was seen only in
recipients who never had an acute rejection episode, emphasizing the recipient's
alloimmune response as a major determinant of overall outcome of the transplant. In addition recent data [2] indicate that even though acute rejection rates
have been drastically reduced, long-term outcome of renal allografts has not.

Furthermore, a transplant recipient must be treated with immunosuppressive agents for life, a therapy that trades the morbidity and mortality of organ failure for the risks of infection and cancer [3]. These drugs are also likely to contribute to increased mortality from cardiovascular disease, the major cause of premature death in kidney transplant recipients [4]. In addition, there is the problem of chronic rejection, which arises at least in part because immunosuppressive strategies do not completely inhibit alloimmune responses and results in a slow progressive deterioration in graft function [5]. These challenges together with the increasing demand of organs for transplantation, create an urgent need for optimizing the outcome of transplanted organs by achieving long-term, drug-free, graft acceptance with normal graft function. Ever since the seminal experiments conducted by Billingham et al. [6] in 1953, there was unequivocal proof of concept that specific tolerance to a defined set of donor antigens can be acquired throughout life. The fact that functional tolerance is not genetically encoded on specific genes has been the fundamental scientific basis for a broad spectrum of investigations to design possible strategies to alter human's immunological response pattern. Achieving the specific goal of donor-specific tolerance would not only minimize the risk of the recipient to suffer from serious side effects resulting from continuous immunosuppressive therapy, but also it would prevent loss of long-term graft function caused by chronic rejection processes, thus making more organs available for primary (first) transplant recipients.

It is, therefore, timely to reassess where we stand on the road to achieving clinical transplant tolerance, and highlight the challenges that face us, so that we may choose the best direction in which to invest our efforts in basic and clinical research [7].

The Concept of Transplantation Tolerance

Recently, numerous insights into the dynamic interrelationship of host immune responses elicited by donor antigen presentation, either on the graft itself or on specialized antigen-presenting cells (APC) have substantially broadened our understanding of the cascade of events that results in the acquisition of tolerance. By definition, tolerance can be described in general terms as a state of unresponsiveness to self or foreign antigens in the absence of immunosuppressive therapy. Inferably, the tolerogenic state of a genetically unrelated organ must be kept in the context of otherwise unrestricted host immune competence to any potential threat jeopardizing the host. Nevertheless, transplant tolerance does not mean complete unresponsiveness of the immune system towards the graft, rather a lack of destructive immune response towards it, in the presence of generalized immune competence.

T cells are the vital elements that orchestrate the alloimmune response and interact with the alloantigens of the graft by the direct and indirect pathways, recognizing the foreign major histocompatibility complex molecules directly on the donor APC and processed donor antigens as peptides on self-APC, respectively [8]. The T cells reacting to their specific antigen can undergo a number of different responses, namely, 'activation' followed by proliferation and 'differentiation' into effector and memory cells, and 'termination'. Physiological termination of the T-cell immune response forms the basis of inducing donor-specific tolerance in clinical transplantation [9]. Several mechanisms, not necessarily mutually exclusive, have been proposed as the basis of transplantation tolerance: deletional mechanisms (actually in the thymus and in the periphery) in which donor reactive T-cell clones are destroyed, and nondeletional/immunoregulatory mechanisms (including anergy, immunedeviation, active suppression/regulation) [10]. Both deletion and immunoregulation are relevant for allograft tolerance but their relative roles differ from experimental model to model [11]. A further possible mechanism of immunological tolerance that is unique to the transplant setting is microchimerism, the persistence of a small number of donor-derived bone marrow (BM) cells in recipients [12]. Microchimerism may be strictly related to and be the inciting mechanism for activating both deletional and nondeletional mechanisms of tolerance.

Central and Peripheral T-Cell Deletion

Studies in experimental animals have indicated that clonal deletion of maturing T lymphocytes may occur centrally in the thymus following donor hematopoietic cell infusion. TCR-transgenic mouse models [13] and Vβ tracking of T cells responding to superantigens presented by donor major histocompatibility complex class II molecules on APC [14] have been used to document the process of central deletion in mixed chimeras. Even in a rat model of kidney transplant tolerance induced by pretransplant intravenous infusion of donor peripheral blood leukocytes or BM cells, under appropriate immunomodulating conditions, intrathymic microchimerism was documented by the presence of donor major histocompatibility complex class II DNA that correlated with graft survival [15]. Together, these data strongly point to intrathymic clonal deletion of donor reactive T cells as one of the major mechanism maintaining tolerance in allogeneic chimeras. However, thymic deletion cannot account for the tolerization of preexisting mature donor-reactive T cells that is achieved in the presence of an intact recipient T-cell repertoire by the use of BM transplant protocols. This observation led to the exploration of peripheral mechanisms through which mature donor reactive T cells are rendered tolerant to donor alloantigens.

Experimental studies of allogeneic BM transplantation with costimulatory blockade in thymectomized recipients have documented clonal deletion of donor-reactive CD4+ T cells, which provides support to the possibility that tolerogenic mechanisms also operate by deletion processes in the periphery [16, 17].

Nondeletional Mechanisms

T cell anergy/regulation are highly complementary with deletion processes and both may well be necessary for long-term transplant tolerance to be achieved. Anergy is a state of functional inactivation in which antigen-specific T lymphocytes are present but are unable to respond. Unresponsiveness can be assessed in vitro by failure of proliferation and cytokine production [18] and in vivo by failure of clonal expansion [19]. Sustained exposure to antigens can also result in the generation of anergic T cells with regulatory capacity that are predominantly CD25−GITR+CTLA-4+/−, even in the absence of tutoring by any pre-existing regulatory T cells [20, 21]. These data support the notion of a form of peripheral tolerance, expounded over a decade ago [22], where anergic T cells can compete out emerging naive responding cells that then default to tolerance themselves.

The role of regulatory cells has been clearly documented in models of tolerance with or without mixed chimeras [15, 23]. Several subsets of regulatory T cells with distinct phenotypes and mechanisms of action have now been identified. They contribute a network of heterogeneous CD4+ or CD8+ T cell subsets and other minor T-cell populations such as nonpolymorphic CD1d-responsive natural killer T cells [23]. Regulatory T cells not only contribute to maintain self-tolerance and prevent autoimmune disease, but can also be induced by tolerance protocols. Although much cellular and molecular characterization has been performed on these cells, there are still many unanswered questions. At the forefront are the following: how do regulatory T cells deliver the suppressive signals? What is the interaction between different regulatory T-cell populations or regulatory dendritic cells (DCs)?

Tolerogenic Strategies in Experimental Animal Models

The use of BM transplantation in order to induce tolerance has been extensively studied in animal models [24]. Establishing mixed chimeric immune systems, with components from the donor and recipient BM, allows tolerance towards the host tissues as well as the foreign graft. A major challenge remains to develop clinically applicable nonmyeloablative regimens that will allow BM transplantation and induction of lasting chimerism, and that can be safely used in HLA-mismatched patients.

An alternative approach to BM chimerism involves the use of in vitro-manipulated or immature donor DCs, which can induce both peripheral and central tolerance [25]. Traditionally, DCs are believed to display properties aimed at sensitizing T lymphocytes specific for foreign antigens. Now, DCs have also become the focus of intense interest as regulators of immune response and it is clear that DC maturation and/or function can be manipulated to promote their tolerogenicity with potential for therapeutic application in organ transplantation.

Recently, significant progress has been made to direct in vitro and in vivo embryonic stem cells (ESCs) to differentiate into specific tissues, including hematopoietic cells. Achievement of mixed chimerism by the use of donor ESC could facilitate the induction of tolerance in the transplant setting. Published evidence suggests that ESCs are immuneprivileged in allogeneic combinations [26]. One of the protective mechanisms against host T-cell mediated rejection could be the constitutional expression of FasL by ESC, but many other factors appear to contribute to this unique property, which could be exploited for the induction of transplantation tolerance. Although the low immunogenicity of ESC may provide an advantage over hematopoietic stem cells, several hurdles, including our little knowledge about the robustness of ESC immune privilege and the underlying mechanisms, remain before clear concepts can be worked out for the use of ESCs in clinical organ transplantation.

Gene therapy is another innovative approach [27]. Elucidating the optimal conditions in nonhuman primates and clarifying the risks associated with such approaches are the first hurdles to be overcome before moving on to clinical trials of these strategies.

Other strategies, utilizing T-cell depleting agents or costimulatory blockade with or without donor-specific transfusion, appear to achieve tolerance in a variety of animal models [9], but not in a true sense of the word in primate models [28]. In the past several years, there has been great excitement about the potential of translating strategies targeting the CD28/CTLA-4:B7–1/2 and the CD40: CD154 T-cell costimulatory pathways to the clinic [8]. Our understanding of these important costimulatory pathways and their interaction with each other and other novel pathways such as ICOS:ICOSL, CD134-CD134L, CD27:CD70, and PD-1:PD-L1/2 are still unfolding. These novel pathways appear to play greater roles under some circumstances [29]. Targeting of these pathways may, however, only work when the alloreactive T-cell repertoire is rendered to a manageable size with adjunctive depleting or deletional therapies [30].

Challenges to Achieving Clinical Transplantation Tolerance

Very small minorities of patients, who discontinue their immunosuppression, provide rare examples of clinical transplantation tolerance [31]. The basis of this immunosuppression-free tolerant state, however, remains intriguing and merits further study so that we may learn how this can be achieved for reproducibility. This phenomenon has also been reported in patients receiving total lymphoid irradiation as induction therapy [32, 33] and in those kidney transplant recipients who had received a previous BM transplant from the same donor [34], and in 2 patients with end-stage renal disease secondary to κ light-chain multiple myeloma who underwent a combined BM and kidney transplantation after a non-myeloablative conditioning regimen [35, 36]. Although effective and appropriate for patients with hematological malignancies, the risks of infection and aplasia and ultimately death associated with the actually available conditioning regimens significantly outweigh the potential benefit of tolerance, particularly when considering the excellent short-term outcomes currently achieved with conventional immunosuppression in organ transplantation. Therefore, some investigators are now attempting to induce donor-specific immune unresponsiveness using donor BM cell infusion without any recipient conditioning [37]. Although all these clinical studies are promising, they do not provide a definite evidence of tolerance induction, since immunosuppressive drugs were not withdrawn at any time post-transplant. So far the only attempt to achieve true tolerance in human renal transplantation with the hematopoietic cell approach has been done using donor G-CSF-mobilized peripheral blood CD34+ stem cells under a nonmyeloablative conditioning regimen of total lymphoid irradiation and anti-thymocyte globulin [38]. Three out of 4 patients achieved multilineage macrochimerism. Maintenance immunosuppression with Cyclosporine A and prednisone was withdrawn in a patient by month 12 post-transplant. In another prednisone was discontinued at month 9 and Cyclosporine A was tapered thereafter. All patients, however, eventually developed some form of rejection and returned to immunosuppressive therapy, although maintenance immunosuppression was considerably lower than conventional regimens. Thus, although these trials represent some progress in the use of donor hematopoietic cells, the goal of stable mixed chimerism resulting in lifelong tolerance remains elusive. More insights about the tolerogenic pathways activated by hematopoietic cell strategy would provide more solid rationale for designing potentially more successful clinical protocols of tolerance induction in organ transplantation.

How can we judge whether such a state has been achieved? Many reports claim tolerance induction after graft survival of more than 100 days in rodents with donor-specific hyporesponsiveness, measured by acceptance of a second graft from the original donor strain and rejection of third-party graft. It is impractical to confirm tolerance induction in this way, in humans. Consequently, devising an assay that allows us to prospectively follow the status of the immune response toward the graft and detect tolerance or early signs of rejection is an urgent necessity [39]. Yet, it seems unlikely that a single assay will provide an adequate immunological profile and a panel of assays may be required. So far, a number of promising assays have emerged but their wider clinical validation is still called for. Such assays should allow us to make decisions regarding prospective withdrawal of immunosuppression without risking a rejection episode in tolerant patients during or after the withdrawal of immunosuppression.

Another major challenge is the definition of the precise impact of the conventional immunosuppressive drugs on tolerizing strategies. Early fears that certain drugs would impair the generation of tolerance have not proven founded in others [40, 41].

The impact of tolerizing regimens on the risk of infectious complications and likewise the detrimental effect of previous, ongoing or later infections on the induction or maintenance of tolerance and also on the course of infection itself is uncertain. Indeed, certain tolerizing strategies are ineffective if performed during ongoing infectious episodes [42]. A recent study has provided more insights on the possible mechanism responsible for this phenomenon, suggesting that individuals harboring virally induced memory T cells that are cross-reactive with donor alloantigens are resistant to tolerance induction [43]. On the other hand, attempting to use a tolerizing regimen in the presence of a latent infectious agent may allow tolerance to develop toward it too. Therefore, it seems prudent to exclude patients with certain chronic or latent infections – such as hepatitis B or C, cytomegalovirus, Epstein-Barr virus – from initial tolerance trials. Nevertheless, the choice of which patient population will be the first to be enrolled into protolerogenic trials is a very difficult one, especially when the clinicians are faced with the ethical issue of risking possible rejection from a failed tolerance protocol when one-year graft survival rates exceed 90% and few grafts are lost to rejection.

Finally, the proper conduct and execution of the clinical trials is a critical issue. The Immune Tolerance Network (ITN), NIH (USA) (*http://www.immune-tolerance.org*) – a consortium of international investigators and research groups dedicated to devising strategies and tools to induce, maintain, and monitor tolerance following islet or solid organ transplantation, as well as in autoimmune and allergic diseases – was expressly instituted for this sole purpose (table 1).

Conclusion

Clinical transplant tolerance is achievable in particular animal models but also in a few humans. Identifying the most successful of these strategies and then translating them to larger animals to test their suitability for the patients is the next step. Although we are currently only at a very early stage, there is no doubt that in the near future some of these approaches will have major impact in transplant

Table 1. Active clinical trials of protolerogenic therapies under the auspices of Immune Tolerance Network (ITN) [44]

Transplant	Therapy
Islet	Daclizumab, sirolimus, and low-dose tacrolimus: Edmonton protocol
Islet	hOKT3-γ-1 (ala-ala) and sirolimus monotherapy
Islet	Campath-1H and one-year temporary sirolimus maintenance monotherapy
Bone marrow and kidney	Nonmyeloablative conditioning regimen (cyclophosphamide, anti-thymocyte globulin, and thymic irradiation) in patients with end-stage renal failure due to multiple myeloma
Bone marrow and kidney	Cyclosporine, cyclophosphamide, MEDI-507, and thymic irradiation
Kidney	Campath 3 combined with sirolimus and mycophenolate mofetil
Kidney	Campath-1H combined with sirolimus and a short course of tacrolimus

medicine, opening a new prospective of indefinite graft survival without the complications of long-term immunosuppressive drugs, and contributing to make a reality donor-specific tolerance in human transplantation.

References

1 Hariharan S, Johnson CP, Bresnahan BA, Taranto SE, McIntosh MJ, Stablein D: Improved graft survival after renal transplantation in the United States, 1988 to 1996. N Engl J Med 2000; 342:605–612.

2 Meier-Kriesche H-U, Schold JD, Srinivas TR, Kaplan B: Lack of improvement in renal allograft survival despite a marked decrease in acute rejection rates over the most recent era. Am J Transplant 2004;4:378–383.

3 Denton MD, Magee CC, Sayegh MH: Immunosuppressive strategies in transplantation. Lancet 1999;353:1083–1091.

4 Kramer BK, Zulke C, Kammerl MC, Schmidt C, Hengstenberg C, Fischereder M, Merienhagen J, European Tacrolimus vs. Cyclosporine Microemulsion Transplantation Study Group: Cardiovascular risk factors and estimated risk of CAD in a randomized trial comparing calcineurin inhibitors in renal transplantation. Am J Transplant 2003;3:982–987.

5 Womer KL, Vella JP, Sayegh MH: Chronic allograft dysfunction: Mechanism and new approaches to therapy. Semin Nephrol 2000;20:126–147.

6 Billingham RE, Brent L, Medawar PB: Actively acquired tolerance of foreign cells. Nature 1953; 172:603–606.

7 Salama AD, Remuzzi G, Harmon WE, Sayegh MH: Challenges to achieving clinical transplantation tolerance. J Clin Invest 2001;108:943–948.

8 Sayegh MH, Turka LA: The role of T-cell costimulatory activation pathways in transplant rejection. N Engl J Med 1998;338:1813–1821.

9 Dong VM, Wormer KL, Sayegh MH: Transplantation tolerance: The concept and its applicability. Pediatr Transplant 1999;3:181–192.

10 Wekerle T, Blaha P, Koporc Z, Bigenzahn S, Pusch M, Muehlbacher F: Mechanism of tolerance induction through the transplantation of donor hematopoietic stem cells: Central versus peripheral tolerance. Transplantation 2003;75:S21–S25.

11 Zheng XX, Sanchez-Fueyo A, Domenig C, Storm TB: The balance of deletion and regulation in allograft tolerance. Immunol Rev 2003;196:75–84.

12 Starzl TE, Demetris AJ, Murase N, Ildstad S, Ricordi C, Trucco M: Cell migration, chimerism, and graft acceptance. Lancet 1992;339:1579–1582.

13 Manilay LO, Pearson DA, Sergio JJ, Swenson KG, Sykes M: Intrathymic deletion of alloreactive T cells in mixed bone marrow chimeras prepared with a nonmyeloablative conditioning regimen. Transplantation 1998;66:96–102.

14 Tomita Y, Kahan A, Sykes M: Role of intrathymic clonal deletion and peripheral anergy in transplantation tolerance induced by bone marrow transplantation in mice conditioned with a non-myeloablative regimen. J Immunol 1994;153:1807–1804.

15 Noris M, Cugini D, Casiraghi F, Azzollini N, De Deus Viera Moraes L, Mister M, Pezzotta A, Cavinato RA, Aiello S, Perico N, Remuzzi G: Thymic microchimerism correlates with the outcome of tolerance-inducing protocols for solid organ transplantation. J Am Soc Nephrol 2001;12:2015–2826.

16 Wekerle T, Kurtz J, Ito H, Ronquillo JV, Dong V, Zhao G, Shaffer J, Sayegh MH, Sykes M: Allogeneic bone marrow transplantation with co-stimulatory blockade induces macrochimerism and tolerance without cytoreductive host treatment. Nat Med 2000;6:464–469.

17 Wekerle T, Sayegh MH, Hill J, Zhao Y, Chandraker A, Zhao G, Sykes M: Extrathymic T cell deletion and allogeneic stem cell engraftment induced with co-stimulatory blockade is followed by central T cell tolerance. J Exp Med 1998;187:2037–2044.

18 Schwartz RH: A cell culture model for T lymphocyte clonal anergy. Science 1990;248:1349–1356.

19 Chen W, Issazadeh S, Sayegh MH, Khoury SJ: In vivo mechanism of acquired thymic tolerance. Cell Immunol 1997;179:165–173.

20 Chen T-C, Cobbold SP, Fairchild PJ, Waldmann H: Generation of anergic and regulatory T cells following prolonged exposure to a harmless antigen. J Immunol 2004;172:5900–5907.

21 Vermeiren J, Ceuppens JL, Van Ghelue M, Witters P, Bullens D, Mages HW, Kroczek RA, Van Gool SW: Human T cell activation by costimulatory signal-deficient allogeneic cells induces inducible costimulator-expressing anergic T cells with regulatory cell activity. J Immunol 2004;172:5371–5378.

22 Waldmann H, Qin S, Cobbold S: Monoclonal antibodies as agents to reinduce tolerance in autoimmunity. J Autoimmun 1992;5(suppl A):93.

23 Wood KJ, Sakaguchi S: Regulatory T cells in transplantation tolerance. Nat Rev Immunol 2003; 3:199–210.

24 Sykes M: Mixed chimerism and transplant tolerance. Immunity 2001;14:417–424.

25 Thomson AW, Lu L: Dendritic cells as regulators of immune reactivity: Implications for transplantation. Transplantation 1999;68:1–8.

26 Zavazava N: Embryonic stem cells and potency to induce transplantation tolerance. Expert Opin Biol Ther 2003;3:5–13.

27 Tomasoni S, Azzollini N, Casiraghi F, Capogrossi MC, Remuzzi G, Benigni A: CTLA4Ig gene transfer prolongs survival and induces donor-specific tolerance in a rat renal allograft. J Am Soc Nephrol 2000;11:747–752.

28 Kirk AD: Transplantation tolerance: A look at the nonhuman primate literature in the light of modern tolerance theories. Crit Rev Immunol 1999;19:349–388.

29 Watts TH, DeBenedette MA: T cell co-stimulatory molecules other than CD28. Curr Opin Immunol 2000;10:57–62.

30 Li XC, Strom TB, Turka LA, Wells AD: T cell death and transplantation tolerance. Immunity 2001;14:407–416.

31 Burlingham WJ, Grailer AP, Fechner JH Jr, Kusaka S, Trucco M, Kocova M, Belzer FO, Sollinger HW: Microchimerism linked to cytotoxic T lymphocyte functional unresponsiveness (clonal anergy) in a tolerant renal transplant recipient. Transplantation 1995;59:1147–1155.

32 Strober S, Benike C, Krishnaswamy S, Engelman EG, Grumet FC: Clinical transplantation tolerance twelve years after prospective withdrawal of immunosuppressive drugs: Studies of chimerism and anti-donor reactivity. Transplantation 2000;69:1549–1554.

33 Strober S, Dhillon M, Schubert M, Holm B, Engelman EG, Benike C, Hoppe R, Sibley R, Myburgh JA, Collins G: Acquired immune tolerance to cadaveric renal allografts. A study of three patients treated with total lymphoid irradiation. N Engl J Med 1989;321:28–33.

34 Sayegh MH, Fine NA, Smith JL, Rennke HG, Milford EL, Tilney NL: Immunologic tolerance to renal allografts after bone marrow transplants from the same donors. Ann Intern Med 1991;114: 954–955.

35 Spitzer TR, Delmonico F, Tolkoff-Rubin N, McAfee S, Sackstein R, Saidman S, Colby C, Sachs DH, Cosimi AB: Combined histocompatibility leukocyte antigen matched donor bone marrow and renal transplantation for multiple myeloma and end stage renal disease: The induction of allograft tolerance through mixed lymphohematopoietic chimerism. Transplantation 1999;68:480–484.

36 Buhler LH, Spitzer TR, Sykes M, Sachs DH, Delmonico F, Tolkoff-Rubin N, Saidman SL, Sackstein R, McAfee S, Dey B, Colby C, Cosimi AB: Induction of kidney allograft tolerance after transient lymphohematopoietic chimerism in patients with multiple myeloma and end-stage renal disease. Transplantation 2002;74:1405–1409.

37 Ciancio G, Miller J, Garcia-Morales RO, Carreno M, Burke GW 3rd, Roth D, Kupin W, Tzakis AG, Ricordi C, Rosen A, Fuller L, Esquenazi V: Six-year clinical effect of donor bone marrow infusion in renal transplant patients. Transplantation 2001;71:827–835.

38 Millan MT, Shizuru JA, Hoffmann P, Dejbakhsh-Jones S, Scandling JD, Grumet FC, Tan JC, Salvatierra O, Hoppe RT, Strober S: Mixed chimerism and immunosuppressive drug withdrawal after HLA-mismatched kidney and hematopoietic progenitor transplantation. Transplantation 2002;73:1386–1391.

39 Strom TB, Suthanthiran M: Prospects and applicability of molecular diagnosis of allograft rejection. Semin Nephrol 2000;20:103–107.

40 Li Y, Li XC, Zheng XX, Wells AD, Turka L, Strom TB: Blocking both signal 1 and signal 2 of T-cell activation prevents apoptosis of alloreactive T cells and induction of peripheral allograft tolerance. Nat Med 1999;5:1298–1302.

41 Sho M, Sandner SE, Najafian N, Salama AD, Dong V, Yamada A, Kishimoto K, Harada H, Schmitt I, Sayegh MH: New insights into the interactions between T-cell costimulatory blockade and conventional immunosuppressive drugs. Ann Surg 2002;236:667–675.

42 Turgeon NA, Iwakoshi NN, Phillips NE, Meyers WC, Welsh RM, Greiner DL, Mordes JP, Rossini AA: Viral infection abrogates CD8(+) T-cell deletion induced by costimulation blockade. J Surg Res 2000;93:63–69.

43 Adams AB, Williams MA, Jones TR, Shirasugi N, Durham MM, Kaech SM, Wherry EJ, Onami T, Lanier JG, Kokko KE, Pearson TC, Ahmed R, Larsen CP: Heterologous immunity provides a potent barrier to transplantation tolerance. J Clin Invest 2003;111:1887–1895.

44 Matthews JB, Ramos E, Bluestone JA: Clinical trials of transplant tolerance: Slow but steady progress. Am J Transplant 2003;3:794–803.

Giuseppe Remuzzi, MD, FRCP
Mario Negri Institute for Pharmacological Research
Via Gavazzeni 11, IT–24125 Bergamo (Italy)
Tel. +39 035 319888, Fax +39 035 319331, E-Mail gremuzzi@marionegri.it

Ronco C, Chiaramonte S, Remuzzi G (eds): Kidney Transplantation: Strategies to Prevent
Organ Rejection. Contrib Nephrol. Basel, Karger, 2005, vol 146, pp 105–120

....

Dendritic Cells, Tolerance and Therapy of Organ Allograft Rejection

Giorgio Raimondi, Angus W. Thomson

Thomas E. Starzl Transplantation Institute and Department of Surgery and
Immunology, University of Pittsburgh, Pittsburgh, Pa., USA

Abstract

Donor dendritic cells (DCs) and those of host origin play key roles in the instigation and
maintenance of immune responses to organ allografts. In the normal steady state, however,
DCs are important for the maintenance of central and peripheral tolerance. Moreover, the
presence of those cells in donor hematopoietic cell infusions may facilitate the induction of
transplant tolerance. Accrual of information regarding DC tolerogenicity has driven the assessment of DC-based therapy of allograft rejection. Pioneering work demonstrating increased
allograft survival after pretransplant infusion of immature donor-derived DC has prompted
the evaluation of several approaches to the generation of DCs with tolerogenic/regulatory
properties. These include: identification of specific culture conditions for propagation of
homogenous populations of immature DCs; pharmacological manipulation of DCs to stabilize
their immature/tolerogenic phenotype; and genetic modification of DCs to impair their
stimulating ability/enhance their tolerogenicity. These approaches have rendered DCs capable
of markedly prolonging experimental allograft (including kidney transplant) survival and
promoting donor-specific tolerance. Recently identified molecular signaling pathways that
play key roles in the outcome of DC-T cell interaction are likely to become novel targets for
manipulation of allograft immunity and for the promotion of transplant tolerance.

Dendritic cells (DCs) present in both organ allografts and recipients can
induce activation of host immune responses, and stimulate rejection by the
direct and indirect pathways of allorecognition, respectively. However, there is
also strong evidence that DCs play a fundamental role in the induction and
maintenance of tolerance. Consequently, a growing number of studies are
aimed at understanding the factors that determine the tolerogenic functions of
DCs in order to render innovative DC-based therapies to control the rejection
response. This mini-review addresses the role of DCs in transplant outcome

and outlines the direction of current research in terms of 'DC therapy' to promote tolerance induction.

DC Immunobiology

DCs are a heterogeneous population of cells (derived from CD34+ stem cells) with morphology (veil-like processes and dendrites) and mobility well suited to their roles in antigen (Ag) capture and processing and in Ag presentation to rare T cells expressing specific receptors that recognize Ag peptides bound to MHC molecules. DCs show distinctive features that allow their classification as 'professional' Ag-presenting cells (APCs): (1) ability to stimulate naïve CD4+ and CD8+ T cells efficiently; (2) capacity to transport Ag from peripheral tissues to T-cell areas of secondary lymphoid organs (where naïve T cells recirculate); (3) ability to 'cross-present' foreign Ags in the context of MHC class I molecules to Ag-specific CD8+ T cells [1].

In the normal steady state (absence of inflammation/'danger'), DCs reside in the interstitial space of most peripheral tissues, including the commonly transplanted organ/tissues, with the exception of the central cornea and brain parenchyma. DCs are abundant in the T-cell areas of spleen and lymph nodes and the medulla of the thymus. The significance of this distribution was clarified with the elaboration of a general paradigm for DC activity (based on studies of DC isolation and reinfusion): DCs acquire Ags in the periphery and migrate to T-cell areas (in spleen via the blood, in lymph nodes via the lymph), where they can either initiate an immune response to the Ags presented or simply die [2]. Death is followed by re-presentation of proteins from the dying DCs by resident DCs in the lymph node. This phenomenon could play an important role in the control of immune activation and tolerance induction, as will be considered below. In transplant models, DCs derived from transplanted allogeneic tissues and carrying donor MHC Ags can be identified in the peripheral lymphoid tissues where they can stimulate T cells directly.

During their lifetime (characterized by considerable turnover), DCs pass through various phenotypical stages (generally referred to as 'maturation states') that correlate with different functional activities and, consequently, different abilities to control the immune response. The DCs in normal blood and nonlymphoid tissue are regarded as 'immature,' – a state characterized by the ability to internalize exogenous Ags (through phagocytosis, macropinocytosis and various receptor-mediated endocytic processes), and to process and load the corresponding peptides onto intracellular MHC molecules, but with only weak ability to stimulate immunity due to low surface expression of MHC and accessory molecules (e.g., CD40, CD80, CD86). These immature DCs are equipped with various surface receptors, the ligation of which starts a signaling pathway, characterized by a fundamental step of nuclear translocation of the gene transcription regulatory factor, nuclear factor (NF)-κB, that initiate DC maturation. These receptors interact specifically with exogenous and endogenous mediators released into the microenvironment during inflammation: (1) Bacterial or viral components (e.g., LPS, CpG, double-stranded RNA) recognized by Toll-like receptors 2, 3, 4, 7 and 9; (2) proinflammatory cytokines (e.g., GM-CSF, IL-1β, TNF-α, IFN-α) and cyclo-oxygenase metabolites (e.g., prostaglandin E2) recognized by their specific counter-receptors; (3) specific ligands, expressed on the surface of 'activated' cells (T cells, platelets and mast cells) that are recognized by molecules of the tumor necrosis factor receptor family on the DC surface (i.e., CD40, tumor necrosis factor receptor, receptor activator of NF-κB). As a result of these interactions, DCs are activated to become 'mature' APC. They down-regulate their endocytic activities, translocate peptide-loaded MHC molecules into the plasma membrane and upregulate surface T cell costimulatory (CD80, CD86, OX40 ligand and inducible costimulator ligand) and intercellular adhesion molecules (CD54 and CD58) necessary for the assembly of the immunological synapse [1]. Furthermore, as they mature, DCs produce proinflammatory cytokines and increase surface expression of the chemokine receptor CCR7 that enables their traffic to T-cell areas of secondary lymphoid tissues, in response to the CCR7 ligands CCL21 and CCL19. Therein, the DCs encounter rare Ag-specific T cells and function as powerful naïve and memory T cell-priming APCs.

DCs and Organ Allograft Rejection

The act of transplantation by itself triggers the maturation and migration of graft-resident and graft-infiltrating DCs [3], where the eliciting stimuli, according to the danger theory model [4], are the 'danger signals' (such as heat shock proteins, uric acid, HMGB1) released in response to inflammation induced by surgical trauma and associated with necrosis due to ischemia/reperfusion injury. As a first consequence of transplant surgery, graft-resident donor DCs migrate as 'passenger' leukocytes to secondary lymphoid tissues of the recipient, where they present donor MHC molecules to recipient T cells via a mechanism known as the 'direct pathway' of allorecognition (donor MHC + peptide X → recipient T cell). This process is characterized by a surprisingly high proportion of circulating T cells (approximately one out of two hundred) [1] that recognize allogeneic MHC molecules. In addition to donor DCs, recipient DCs or DC precursors, mobilized to the graft as part of the initial inflammatory infiltrate, acquire donor allo-Ag (by internalization of soluble MHC molecules, fragments/blebs derived from donor apoptotic or necrotic cells, or by vesicle and possibly exosome [5] exchange between living cells) and present MHC-derived peptides bound to self-MHC molecules to recipient T cells. This phenomenon is known as the 'indirect pathway' of allorecognition (self-MHC + donor MHC-derived peptide → recipient T cell). The relative importance of these two pathways in the rejection response has been the subject of controversy for several years [6]. Classical experiments have shown that there is an increase in the graft survival when thyroid, pancreatic islet, skin, and kidney allografts are purged of interstitial leukocytes [1]. These observations provided a basis for the concept that the direct pathway of allorecognition is the most important component of the acute graft rejection response. More recently, it has been demonstrated that the role of the direct versus the indirect pathway in graft rejection depends on the type of organ/tissue transplanted, the experimental model, and the phase of rejection [3, 7]. Both pathways participate in the early phases of acute rejection, but several clinical observations indicate that T-cell responses elicited by the direct pathway decrease with time after transplantation [3], whereas the role of the indirect pathway is sustained and participates in chronic rejection [3]. It has been argued that initial and repeated immune-mediated damage caused during acute rejection can predispose to chronic rejection [8]. Recognition of the role of both pathways in events leading to rejection justifies targeting both direct and indirect allorecognition in strategies to promote organ transplant tolerance.

DC Subsets

Any therapy aimed at controlling the alloimmune response by targeting DCs or by using adoptively transferred DCs to promote tolerance (see below) has to confront the heterogeneity of DC populations that exhibit distinctive features, in terms of functional significance and immune-modulating potential. Recent reviews [9, 10] have covered this topic in depth, addressing in detail the relationships between different DC subsets in mice and humans.

In the spleens of normal mice, two main populations of CD11c+ DCs can be identified based on the expression of the CD8αα homodimer and the β2 integrin CD11b/CD18: CD8α−CD11b+CD4+CD205− DCs (originally referred to as 'myeloid' DCs; MDC), located predominantly in the splenic marginal zone, and CD8α+CD11b−CD4−CD205+ (originally referred to as 'lymphoid-related' DCs), that concentrate in the T cell-dependent areas of the splenic follicle. Lymph node DCs also include two other CD8α−CD11b+ DC subpopulations that migrated from peripheral tissues: CD205+ and CD205− DCs, probably derived from peripheral tissue-resident or interstitial DCs, and Langerin+ DCs, derived from migratory epidermal Langerhans cells. The mouse thymic medulla is populated exclusively by CD8α+ DCs that are involved in the positive/negative selection of autoreactive thymocytes. CD8α+ DC and CD8α− MDC appear to differentially regulate Th cell responses [1]. It was suggested initially that, due to their in vitro functional properties, murine CD8α+ might be DC specialized for tolerance induction [1], but other findings conflict with this view. Moreover, there is no phenotypic counterpart of murine CD8α+ DC in humans.

Two main populations of DC have been described in humans, where knowledge of their functional biology is more limited. Besides the classic MDC (generated from circulating monocytes or CD34+ hematopoietic stem cells) a second subset of DC termed 'plasmacytoid DC' (pDCs) has been found recently in the circulation and in T-cell areas of secondary lymphoid tissue [11]. pDCs are identical to previously described natural type-1 IFN-producing cells that produce IFN α/β in response to viral activation. It follows that human MDC induce Th1 cell differentiation, whereas human pDCs can selectively induce Th2 cells when stimulated with IL-3 and CD40L or promote Th1 cell differentiation when activated by virus [12, 13]. The mouse counterpart of human pDC has been identified recently.

DCs and Tolerance Induction

DCs have been believed for many years to be involved in negative selection of thymocytes in central tolerance [14]. More recently there is evidence that presentation of peripherally-derived Ags by DCs within secondary lymphoid tissue is crucial for the induction of T-cell tolerance to self-Ags expressed exclusively in peripheral tissues [15]. Different models have been proposed to explain the mechanism(s) by which DCs may induce/maintain peripheral T-cell tolerance [16]. Steinman et al. [15] have proposed that under steady-state conditions (no inflammation), the uptake of Ags by immature DCs expressing low cell surface levels of MHC and costimulatory molecules may induce tolerance to those peptides presented to Ag-specific T cells. This prediction is based on two experimental observations: (1) Binding of the T cell receptor on naïve T cells to MHC-peptide complexes on the APC in the absence of or with low costimulation leads to Ag-specific T-cell unresponsiveness; (2) in the healthy steady state, DCs continuously traffic from the periphery to secondary lymphoid tissue transporting self-Ags [16]. However, the concept of migratory immature DCs as the keepers of peripheral T-cell tolerance disagrees with the observation that lymph-borne DCs, obtained by cannulation of lymphatic vessels in the steady state, exhibit signs of maturation in all animal models investigated so far [1]. This discrepancy may be explained by the fact that a certain degree of DC maturation is required for homeostatic DC trafficking in the absence of inflammation. This state has been defined as 'semi-mature' by Lutz and Schuler [17]. At this 'semi-mature' stage, DCs could express levels of costimulatory molecules and produce levels of proinflammatory cytokines insufficient to activate an immune response and instead induce tolerance. Obviously, these concepts would also be compatible with different models that outline the capacity of the immune system to perceive evolving situations and respond appropriately, depending not only on the activation state of the APCs but also on the level of Ag presented and the persistence of Ag presentation [18, 19]. Independent of the validity of one model versus another, different groups have shown using transgenic mice, that constitutive migratory MHC II+ APCs that transport tissue-specific Ags from the periphery silence, rather than activate, Ag-specific CD4+ or CD8+ T lymphocytes in secondary lymphoid tissues [1].

The precise mechanism(s) by which immature or semi-mature DCs induce specific T-cell tolerance to self or non-self-Ags is not well understood, and current evidence suggests that more than one mechanism may be involved. First, immune deviation, or skewing of T cells toward the Th2 type, appears to be a mechanism that DCs exploit. Thus, several groups have shown that DCs can induce immune deviation in autoimmune disease and transplant models [20]. This effect seems to be enhanced by costimulation blockade using the fusion protein 'cytotoxic T lymphocyte Ag 4' (CTLA4)-Ig, that blocks the B7-CD28 signaling pathway [21], during Ag loading or treatment of DCs in culture (i.e., with IL-10) in order to impair their Th1-promoting activity and increase Th2 skewing [22, 23]. A second mechanism underlying tolerance is the induction of anergy (a state of T-cell unresponsiveness, reversible under specific conditions) [24] or apoptosis. Many studies have shown that MDC whose allostimulatory function is impaired, either by incomplete maturation, selective blockade of B7 costimulatory molecules, the influence of specific cytokines (e.g., IL-10 or TGF-β), or genetic engineering (to express viral IL-10, CTLA4-Ig, or FasL), can induce alloAg-specific T-cell hyporesponsiveness (anergy) or apoptosis in vitro, and suppress immune reactivity [20, 25]. In particular, considering the perception of Ag-specific T-cell deletion as a robust tolerance-inducing mechanism, various reports have shown that overexpression of molecules associated with the induction of apoptosis, i.e., FasL [26, 27], nitric oxide [28, 29] or the tryptophan-catabolizing enzyme IDO [30] may render DC capable of subverting T-cell responses by promoting activation-induced cell death. Blockade of the B7/CD28 pathway by CTLA4Ig significantly increases MDC-induced apoptosis of alloactivated T cells [27]. This appears to be mediated, at least in part, via the Fas pathway. On the other hand, a recent investigation indicates that ligation of B7 molecules by CTLA4-Ig induces up-regulation of IDO production in DC causing local reduction in tryptophan availability and the presence of proapoptotic kynurenines that decrease clonal expansion and enhance T-cell deletion [30]. Third, evidence has emerged that DCs can promote the induction of cells with regulatory activity, a mechanism investigated recently by a large number of laboratories due to its potential to ensure long-term, Ag-specific unrensponsiveness. Various T-regulatory (Treg) cell populations, especially CD4+ T cells, that exhibit suppressor activity, have been described: Tr1 and Th3 cells that are induced following T-cell activation in the presence of IL-10 and/or TGF-β, CD4+CD25+ cells that arise spontaneously during ontogeny and are present in the periphery of normal mice, and also CD8+CD28−, CD3+CD4−CD8− and NKT cells with regulatory activity have been described after transplantation [31]. Interest in the generation of Treg cells has recently been paralleled by the emerging concept that interaction between tolerogenic DCs (also termed regulatory DCs) and Treg cells may constitute an inhibitory feedback loop that can prevent organ allograft rejection [1].

The majority of data that support the role of DCs in controlling immune responses have been generated in animal models, but there are many clear indications from human studies of the regulatory potential of DC-based therapeutic strategies. Human CD4+ T cells with characteristics of Treg cells (low proliferative capacity, secretion of IL-10, and ability to inhibit alloAg-specific proliferation of other T cells) can be generated in vitro following repetitive stimulation of naïve CD4+ T cells with allogeneic immature DCs [32]. Furthermore, in human volunteers, immature autologous monocyte-derived DCs pulsed with the human leukocyte Ag-A*0201-restricted influenza matrix peptide induced specific inhibition of MP-specific CD8+ cytotoxic T lymphocytes and induced IL-10-secreting CD8+ T lymphocytes [33]. These results constitute proof of principle of DC tolerogenicity in humans that is already driving new investigations in DC-based therapy that are further investigating the importance of tolerogenic DCs-T cell interaction in regulation of alloimmune responses and also the various parameters that may influence this interaction (e.g., the route of DC in vivo administration, the temporal relationship between their administration and that of Ag, the number of cells injected, the amount and physical condition of Ag presented, and the source, type and maturation state of the DC).

DCs and the Control of Organ Transplant Outcome

At face value, the contribution of DCs to transplant rejection may seem quite straightforward within the framework of direct and indirect allorecognition. However, in apparent contradiction to the classical passenger leukocyte experiments, several groups have reported that depletion of donor bone marrow

(BM)-derived cells prevents the induction of transplantation tolerance [3, 34]. For example, depletion of passenger leukocytes from rat donor heart allografts reversed the beneficial effects of donor-specific blood transfusion. Furthermore, tolerance was re-established if donor-type DCs were cotransferred at the time of transplantation of APC-depleted cardiac grafts [34]. Another set of data that relate to the role of donor DCs in transplant immunity concerns the phenomenon of donor hematopoietic cell 'microchimerism' as observed by Starzl and coworkers [35, 36]. They detected donor hematopoietic cells in lymphoid and nonlymphoid tissues of long-surviving, successful human organ allograft recipients. It was proposed that the ability of an organ to be tolerogenic, in the absence or presence of effective immunosuppression, was dependent on its passenger leukocyte, and not its parenchymal cell component [35]. In support of this view, the radiosensitive passenger leukocyte population of renal allografts has been implicated in the induction of tolerance to contemporaneous heart grafts in a miniature swain model [37]. Even if some issues remain unresolved regarding this hypothesis [3], the general understanding of the role of DCs in peripheral tolerance, together with the development of techniques to generate large numbers of DCs in vitro, have opened up the possibility of generating DCs with tolerogenic/regulatory properties for therapeutic application.

Various experimental techniques have been used to generate DC progenitors or precursors or DCs with tolerogenic potential. These can be grouped under three technological approaches: (1) Specific culture conditions, (2) pharmacological manipulation, and (3) genetic engineering.

DC Manipulation for Tolerance Induction

In line with the classic model of T-cell stimulation dependent on the maturation state of the APC, the generation of immature DCs (MHC+, $CD80^{lo/-}$, $CD86^{lo/-}$) in culture was the first attempt to produce tolerogenic DCs [38]. This early work showed that immature MDC could induce alloAg-specific T-cell hyporesponsiveness in vitro. Subsequently, various studies have shown that, if administered before, during, or even after transplantation, immature donor DCs can prolong allograft (including skin graft) survival. In some instances, indefinite, donor-specific graft survival is achieved [39–43]. One of the potential drawbacks of this approach is the possibility that, following injection, a fraction of the immature donor DCs differentiate in vivo into mature APCs, with the ability to stimulate an anti-donor response and accelerate graft rejection. In an effort to overcome this potential problem, some investigators have combined administration of immature, donor-derived DCs with a short course of anti-CD40L (anti-CD154) mAb (theoretically to avoid potential maturation of the injected DCs in the recipient) obtaining striking enhancement of graft survival [1]. Administration of LPS-, TNF-α-, and CD40-maturation resistant donor-derived DCs has proven to be an alternative means to avoid in vivo DC maturation, with indefinite prolongation of heart allograft survival (>100 days) in nonimmunosuppressed mice [44]. Recently, Sato et al. [45] found that mouse BM-derived DCs generated with IL-10, TGF-β and LPS in addition to GM-CSF, acquired regulatory functions; if generated from host BM, they protect mice from lethal, allogeneic BM-induced graft-versus-host disease. Furthermore, these DC induce Ag-specific CD4+CD25+CD152+ Treg cells in the transplant recipients. A similar regulatory effect (induction of anergic and Treg cells) has been shown using human monocyte-derived DC cultured with IL-10 and TGF-β (in addition to GM-CSF and IL-4) [46].

In an effort to obtain DCs with a stable, immature phenotype, or with impaired ability to synthesize Th1-driving cytokines (i.e., IL-12p70), DCs have been treated with various pharmacological agents. The spectrum of molecules investigated includes: aspirin, cyclic adenosine monophosphate inducers (prostaglandin E2, histamine, β2 agonists, neuropeptides), the vitamin D3 metabolite $1\alpha,25$-$(OH)_2D3$ and its analogs, glucosamine, the antioxidant N-acetyl-L-cysteine and immunosuppressive drugs (corticosteroids, cyclosporine A, rapamycin, deoxyspergualin, and mycophenolate mofetil) all shown to prevent DC activation/maturation or to impair the capacity of DCs to produce bioactive IL-12p70 in vitro and in vivo [47]. Of particular interest, treatment with donor-derived DCs generated in vitro in the presence of the active form of vitamin D3, $1\alpha,25$-$(OH)_2D3$, in combination with mycophenolate mofetil, induces tolerance to fully mismatched mouse pancreatic islet allografts [48] with generation/amplification of Treg cells (able to confer protection against islet rejection in naïve animals). Parallel to this is the treatment of DCs with immunosuppressive drugs, in particular dexamethasone that arrests their differentiation/maturation [49] and offers potential for development of 'negative cellular vaccines' for immunotherapy.

Recent advances in gene transfer technology have resulted in enhancement/ stabilization of the tolerogenic potential of DCs following their genetic modification to express 'immunosuppressive' molecules that can (1) inhibit or block cell-surface costimulatory molecule expression (IL-10, TGF-β, CTLA4-Ig), (2) prevent proliferation of allogeneic T cells (IDO), (3) induce and maintain T-cell anergy (B7-H1), (4) promote the deletion of Ag-specific T cells (CD95L, TRAIL) [25]. To date, there have been no reports of donor-specific tolerance being achieved across MHC barriers using genetically modified, donor-derived DCs alone. However, of considerable significance is the finding that a single, pretransplant infusion of NF-κB ODN (NF-κB specific 'decoy' oligodeoxyribonucleotides)-treated, CTLA4-Ig-transduced donor MDCs markedly prolong fully MHC-mismatched vascularized heart allograft survival, with 40% of the animals exhibiting long-term (>100 day) graft survival [50].

All these approaches have been tested using MDCs. Identification of additional DC subsets with different functional properties has led to their investigation for tolerogenic potential. In particular, considerable interest has grown around plasmacytoid (p)DCs. It has been reported [1] that immature pDCs, freshly isolated from human peripheral blood, can induce Ag-specific anergy in CD4+ T cell lines. This phenomenon may involve the inhibitory receptors Ig-like transcript 3 and 4, expressed on the surface of immature pDCs [1]. Furthermore, in mice, our preliminary data indicate that a single pretransplant infusion of highly purified, freshly isolated donor pre-pDCs from mouse secondary lymphoid tissue markedly prolongs vascularized organ allograft survival [51]. Compared to results obtained with MDCs and CD8α+ DCs in mice under the same experimental conditions, pDCs exhibit a more pronounced tolerogenic effect.

DC Therapy – The Indirect Pathway

The impressive results obtained with various strategies of donor-derived DC treatment in small animal models seem to contrast with the previously described concept of rejection determined by both direct and indirect pathways of T-cell activation and the need to inhibit both to properly control the rejection response. However, it has been shown that donor-derived DCs can transfer allogeneic MHC molecules to recipient DCs in vivo [3]. Thus, it has been argued that administration of donor DCs may also exert an influence on the indirect pathway of allorecognition. Given that the role of the direct pathway diminishes with time after transplantation, while that of the indirect pathway appears to be sustained, and participates in chronic rejection, attempts have been made to target the indirect pathway via DC therapy. In a rat model, recipient DCs (BM-derived or thymic) pulsed ex vivo with immunodominant donor MHC I-derived allopeptides, were injected into the thymi of recipients, 7 days before transplant. This led to permanent survival of cardiac or islet grafts when administered with antilymphocyte serum [52, 53]. A similar effect was observed when recipient DCs, pulsed with donor allopeptides, were administered intravenously – a more feasible route in relation to possible clinical application [54, 55].

This strategy of ex vivo pulsing of recipient DCs with immunodominant MHC-derived peptides (or other sources of donor MHC, such as apoptotic bodies or exosomes), is likely to yield improved results with use of pharmacologically modified DCs. A promising example concerns preoperative injection of mouse organ allograft recipients with rapamycin-treated, donor alloAg-pulsed, recipient-derived DCs that significantly prolongs graft survival in a donor-specific manner [56]. Use of specific (more tolerogenic) recipient DC subsets may give even better results, considering the observation that pre-pDC represent a significantly higher proportion of circulating DC precursors (compared to MDC) in tolerant human organ transplant recipients, compared with patients requiring maintenance immunosuppression [57]. As an alternative, genetic engineering of recipient DCs may also be considered. Billing et al. [58] have shown that administration of immature recipient DCs encoding a donor MHC I transgene prolongs cardiac allograft survival.

DC therapy also seems to be effective in preventing/ameliorating chronic rejection (transplant vascular sclerosis). Thus, Wang et al. [59] have demonstrated that peritransplant administration of purified donor immature splenic DCs, in combination with blocking anti-CD154 mAb, strongly inhibits the development of intimal thickening, fibrosis, and proliferation of α-smooth muscle actin+ cells in a murine aortic allograft model.

DC Therapy and the Outcome of Kidney Transplantation

Various experimental transplant models have validated the idea of targeting DCs to prevent kidney allograft rejection and induce a robust form of tolerance (table 1). Impressive results have been obtained in a rhesus macaque renal allograft model, using a combination of peritransplant treatment with a T cell-depleting agent (anti-CD3 immunotoxin) and a 15-day course of deoxyspergualin – an NF-κB inhibitor that suppresses DC maturation and proinflammatory cytokine production [60]. 87% of the recipients showed allograft survival with no form of rejection without the need for continued immunosuppressive treatment. The efficacy of this treatment correlates with significant reduction of mature DCs in recipient lymph nodes, together with the coincident reduction in lymph node T-cell mass. These data provide a striking example of renal transplant tolerance linked to in situ inhibition of DC maturation.

Significant prolongation of mouse renal allograft survival has been achieved by pretransplant portal venous infusion of a mixture of donor-derived MDCs transduced to express TGF-β1 or IL-10. This protective effect correlated with the inhibition of cytotoxic T lymphocyte induction and with enhancement of Th2 responses [61]. Recently, further encouraging and more clinically relevant result has been described by Mirenda et al. [62]. They have shown that pretreatment of rats (7 days before transplant) with dexamethasone-treated DCs (obtained from F1 donors) coexpressing donor and recipient MHC molecules, together with a single dose of CTLA4-Ig (one day later), lead to indefinite kidney allograft survival after a short postoperative course of cyclosporine A (to inhibit the early direct pathway response). This striking therapeutic effect seems to be associated with the presence and function of indirect pathway Treg cells [62], confirming the validity and importance of interventions aimed

Table 1. Dendritic cells, therapy of kidney allograft rejection and tolerance induction

Model	Approach	Result	Comments	References
Rhesus macaque	Peritransplant administration of anti-CD3 immunotoxin + 15 days course of deoxyspergualin	Allografts survived without rejection in 87% of the recipients	Marked reduction in mature DCs in recipient lymph nodes, attributed to deoxyspergualin	[60]
Mouse	Portal vein infusion of a mixture of donor-derived MDCs transduced to express TGF-β1 or IL-10	Consistent extension of graft survival time	The first indication of the efficacy of genetically modified DC in controlling kidney allograft rejection	[61]
Rat	Pretransplant infusion of dexamethasone-treated DCs coexpressing donor and recipient MHC molecules + single dose of CTLA4-Ig + short course of cyclosporine A	Indefinite allograft survival	Tolerance associated with induction of indirect pathway Treg cells	[62]

at modulating the Ag-presenting function of DCs to obtain a robust form of tolerance.

There is recent evidence that in vivo-mobilized kidney DCs are functionally immature and can prolong allograft survival in a mouse organ transplant model [63]. This observation raises interesting questions about the potential tolerogenicity of immature DCs mobilized in potential kidney allograft donors with hematopoietic growth factors.

Conclusion

After more than 20 years of research on their role in organ transplantation outcome, a better understanding of the fundamental role of DCs in regulation of the alloimmune response has been obtained, together with a clear indication that these cells have tolerogenic potential. The evaluation of DC-based therapy for the control of the alloimmune response and the induction of tolerance is in its infancy. Further insights are necessary regarding the elaborate cross-talk between DCs and the effector cells of the adaptive immune response. An example of this emerging knowledge is the recent demonstration of a relationship between expression of different Notch ligands on DCs and their ability to induce Th1/Th2 responses [64], a process that needs to be further understood in order to create a firm basis for better manipulation/control of the alloimmune response. Furthermore, the early 'simple' view of costimulation has recently become more complex by the discovery of several new signaling pathways involving CD70, OX40L and inducible costimulator ligand on DCs whose specific role and activities are currently under investigation. Moreover, an even more complex situation is evident from the identification of novel 'coregulatory' molecules of the B7 family (B7-H1, B7-DC, B7-H3 and B7-H4) that show potent inhibitory effects on T-cell activation, but whose mechanisms of action require further investigation (some of their cognate receptors have not yet been identified), as stimulatory effects have also been described for these molecules [65]. Recent investigations have underscored the potential exploitation of these new pathways to control the alloimmune response [66, 67].

On the whole, these pathways offer new opportunities for targeting/manipulating DC function to modify their ability to regulate immune reactivity. Combination of DC therapy (using specific manipulated DC subsets) with peri- or post-transplant therapies, that target DC-T cell interaction/T cell function, may predispose to stable allograft tolerance and allow weaning of chronic immunosuppressive regimes.

Acknowledgment

The authors' work is supported by National Institute of Health grants RO1 49745, RO1 AI 41011 and UO1 AI 51698.

References

1   Morelli AE, Thomson AW: Dendritic cells: Regulators of alloimmunity and opportunities for tolerance induction. Immunol Rev 2003;196:125–146.
2   Steinman RM: The dendritic cell system and its role in immunogenicity. Annu Rev Immunol 1991;9:271–296.
3   Lechler R, Ng WF, Steinman RM: Dendritic cells in transplantation – Friend or foe? Immunity 2001;14:357–368.
4   Matzinger P: The danger model: A renewed sense of self. Science 2002;296:301–305.

5 Peche H, Heslan M, Usal C, Amigorena S, Cuturi MC: Presentation of donor major histocompat-ibility complex antigens by bone marrow dendritic cell-derived exosomes modulates allograft rejection. Transplantation 2003;76:1503–1510.

6 Gould DS, Auchincloss H Jr: Direct and indirect recognition: The role of MHC antigens in graft rejection. Immunol Today 1999;20:77–82.

7 Illigens BM, et al: The relative contribution of direct and indirect antigen recognition pathways to the alloresponse and graft rejection depends upon the nature of the transplant. Hum Immunol 2002;63:912–925.

8 Libby P: Transplantation-associated arterosclerosis: Potential mechanisms; in Tilney NL, Strom TB, Paul LC (eds): Transplantation Biology: Cellular and Molecular Aspects. Philadelphia, Lippencott-Raven, 1996, p 577.

9 Shortman K, Liu Y-J: Mouse and human dendritic cell subtypes. Nat Rev 2002;2:151–161.

10 Ardavin C: Origin, precursors and differentiation of mouse dendritic cells. Nat Rev Immunol 2003;3:582–590.

11 Siegal FP, et al: The nature of the principal type 1 interferon-producing cells in human blood. Science 1999;284:1835–1837.

12 Rissoan MC, Soumelis V, Kadowaki N, Grouard G, Briere F, de Waal Malefyt R, Liu YJ: Reciprocal control of T helper cell and dendritic cell differentiation. Science 1999;283: 1183–1186.

13 Cella M, Facchetti F, Lanzavecchia A, Colonna M: Plasmacytoid dendritic cells activated by influenza virus and CD40L drive a potent TH1 polarization. Nat Immunol 2000;1:305–310.

14 Matzinger P, Guerder S: Does T-cell tolerance require a dedicated antigen presenting cell? Nature 1989;338:74–76.

15 Steinman RM, Hawiger D, Nussenzweig MC: Tolerogenic dendritic cells. Annu Rev Immunol 2003;21:685–711.

16 Morelli AE, Hackstein H, Thomson AW: Potential of tolerogenic dendritic cells for transplantation. Semin Immunol 2001;13:323–335.

17 Lutz MB, Schuler G: Immature, semi-mature and fully mature dendritic cells: Which signals induce tolerance or immunity? Trends Immunol 2002;23:445–449.

18 Langman RE, Cohn M: A minimal model for the self-nonself discrimination: A return to the basics. Semin Immunol 2000;12:189–195; discussion 257–344.

19 Grossman Z, Min B, Meier-Schellersheim M, Paul WE: Concomitant regulation of T-cell activation and homeostasis. Nat Rev Immunol 2004;4:387–395.

20 Lu L, Thomson AW: Dendritic cell tolerogenicity and prospects for dendritic cell-based therapy of allograft rejection and autoimmune disease; in Lotze MT, Thomson AW (eds): Dendritic Cells, ed 2. San Diego, Academic Press, 2001, pp 587–607.

21 Khoury SJ, Gallon L, Verburg RR, Chandraker A, Peach R, Linsley PS, Turka LA, Hancock WW, Sayegh MH: Ex vivo treatment of antigen-presenting cells with CTLA4Ig and encephalitogenic peptide prevents experimental autoimmune encephalomyelitis in the Lewis rat. J Immunol 1996; 157:3700–3705.

22 De Smedt T, van Mechelen M, De Becker G, Urbain J, Leo O, Moser M: Effect of interleukin-10 on dendritic cell maturation and function. Eur J Immunol 1997;7:1229–1235.

23 Kalinski P, Hilkens CM, Snijders A, Snijdewint FG, Kapsenberg ML: IL-12-deficient dendritic cells, generated in the presence of prostaglandin E2, promote type 2 cytokine production in maturing human naive T helper cells. J Immunol 1997;159:28–35.

24 Schwartz RH: T cell anergy. Annu Rev Immunol 2003;21:305–334.

25 Lu L, Thomson AW: Genetic engineering of dendritic cells to enhance their tolerogenic potential. Graft 2002;5:308–315.

26 Suss G, Shortman K: A subclass of dendritic cells kills CD4 T cells via Fas/Fas-ligand-induced apoptosis. J Exp Med 1996;183:1789–1796.

27 Lu L, Qian S, Hershberger PA, Rudert WA, Lynch DH, Thomson AW: Fas ligand (CD95L) and B7 expression on dendritic cells provide counter-regulatory signals for T cell survival and proliferation. J Immunol 1997;158:5676–5684.

28 Lu L, Bonham CA, Chambers FG, Watkins SC, Hoffman RA, Simmons RL, Thomson AW: Induction of nitric oxide synthase in mouse dendritic cells by IFN-gamma, endotoxin, and interaction with

allogeneic T cells: Nitric oxide production is associated with dendritic cell apoptosis. J Immunol 1996;157:3577–3586.

29 Fehsel K, Kroncke KD, Meyer KL, Huber H, Wahn V, Kolb-Bachofen V: Nitric oxide induces apoptosis in mouse thymocytes. J Immunol 1995;155:2858–2865.

30 Fallarino F, Grohmann U, Vacca C, Orabona C, Spreca A, Fioretti MC, Puccetti P: T cell apoptosis by kynurenines. Adv Exp Med Biol 2003;527:183–190.

31 Wood KJ, Sakaguchi S: Regulatory T cells in transplantation tolerance. Nat Rev Immunol 2003;3: 199–210.

32 Jonuliet H, Schmitt E, Schuler G, Knop J, Enk AH: Induction of interleukin 10-producing, non-proliferating CD4þ T cells with regulatory properties by repetitive stimulation with allogeneic immature human dendritic cells. J Exp Med 2000;192:1213–1222.

33 Dhodapkar MV, Steinman RM, Krasovsky J, Munz C, Bhardwaj N: Antigen-specific inhibition of effector T cell function in humans after injection of immature dendritic cells. J Exp Med 2001; 193:233–238.

34 Josien R, Heslan M, Brouard S, Soulillou JP, Cuturi MC: Critical requirement for graft passenger leukocytes in allograft tolerance induced by donor blood transfusion. Blood 1998;92:4539–4544.

35 Starzl TE, Demetris AJ, Murase N, Ildstad S, Ricordi C, Trucco M: Cell migration, chimerism, and graft acceptance. Lancet 1992;339:1579–1582.

36 Starzl T, Demetris A, Trucco M, Murase N, Ricordi C, Ildstad S, Ramos H, Todo S, Tzakis A, Fung J, Nalesnik M, Zeevi A, Rudert W, Kocova M: Cell migration and chimerism after whole-organ transplantation: The basis of graft acceptance. Hepatology 1993;17:1127–1152.

37 Mezrich JD, Yamada K, Lee RS, Mawulawde K, Benjamin LC, Schwarze ML, Maloney ME, Amoah HC, Houser SL, Sachs DH, Madsen JC: Induction of tolerance to heart transplants by simultaneous cotransplantation of donor kidneys may depend on a radiation-sensitive renal-cell population. Transplantation 2003;76:625–631.

38 Lu L, McCaslin D, Starzl TE, Thomson AW: Bone marrow-derived dendritic cell progenitors (NLDC 145+, MHC class II+, B7–1dim, B7–2-) induce alloantigen-specific hyporesponsiveness in murine T lymphocytes. Transplantation 1995;60:1539–1545.

39 Fu F, Li Y, Qian S, Lu L, Chambers F, Starzl TE, Fung JJ, Thomson AW: Costimulatory molecule-deficient dendritic cell progenitors (MHC class II+, CD80dim, CD86-) prolong cardiac allograft survival in nonimmunosuppressed recipients. Transplantation 1996;62:659–665.

40 O'Connell PJ, Li W, Wang Z, Specht SM, Logar AJ, Thomson AW: Immature and mature CD8alpha+ dendritic cells prolong the survival of vascularized heart allografts. J Immunol 2002;168:143–154.

41 Rastellini C, Lu L, Ricordi C, Starzl TE, Rao AS, Thomson AW: Granulocyte/macrophage colony-stimulating factor-stimulated hepatic dendritic cell progenitors prolong pancreatic islet allograft survival. Transplantation 1995;60:1366–1370.

42 Suri RA, Niimi M, Wood KJ, Austyn JM: Immature dendritic cell pretreatment induces cardiac allograft prolongation. J Leukoc Biol 1998;32:#B16.

43 Hayamizu K, Huie P, Sibley RK, Strober S: Monocyte-derived dendritic cell precursors facilitate tolerance to heart allografts after total lymphoid irradiation. Transplantation 1998;66:1285–1291.

44 Lutz MB, et al: Immature dendritic cells generated with low doses of GM-CSF in the absence of IL-4 are maturation resistant and prolong allograft survival in vivo. Eur J Immunol 2000;30: 1813–1822.

45 Sato K, Yamashita N, et al: Regulatory dendritic cells protect mice from murine acute graft-versus-host disease and leukemia relapse. Immunity 2003;18:367–379.

46 Sato K, Yamashita N, Baba M, et al: Modified myeloid dendritic cells act as regulatory dendritic cells to induce anergic and regulatory T cells. Blood 2003;101:3581–3589.

47 Abe M, Hackstein H, Thomson AW: Manipulation of dendritic cells in organ transplantation: A major step toward graft tolerance? Curr Opin Organ Transplant 2004; in press.

48 Gregori S, Casorati M, Amuchastegui S, Smiroldo S, Davalli AM, Adorini L: Regulatory T cells induced by 1-alpha,25-dihydroxyvitamin D3 and mycophenolate mofetil treatment mediate transplantation tolerance. J Immunol 2001;167:1945–1953.

49 Hackstein H, Thomson AW: Dendritic cells: Emerging pharmacological targets of immunosup-pressive drugs. Nat Rev Immunol 2004;4:24–34.

50 Bonham CA, et al: Marked prolongation of cardiac allograft survival by dendritic cells genetically engineered with NF-κB oligodeoxyribonucleotide decoys and adenoviral vectors encoding CTLA4-Ig. J Immunol 2002;169:3382–3391.

51 Coates PTH, Duncan FJ, Wang Z, Thomson AW, Bjorck P: Plasmacytoid dendritic cells markedly prolong allograft survival in the absence of systemic immunosuppression (abstract). Am J Transplant 2003;3:163.

52 Garrovillo M, Ali A, Oluwole SF: Indirect allorecognition in acquired thymic tolerance: Induction of donor-specific tolerance to rat cardiac allografts by allopeptide-pulsed host dendritic cells. Transplantation 1999;68:1827–1834.

53 Ali A, Garrovillo M, Jin MX, Hardy MA, Oluwole SF: Major histocompatibility complex class I peptide-pulsed host dendritic cells induce antigen-specific acquired thymic tolerance to islet cells. Transplantation 2000;69:221–226.

54 Garrovillo M, et al: Induction of transplant tolerance with immunodominant allopeptide-pulsed host lymphoid and myeloid dendritic cells. Am J Transplant 2001;1:129–137.

55 Oluwole SF, et al: Indirect allorecognition in acquired thymic tolerance: Induction of donor-specific permanent acceptance of rat islets by adoptive transfer of allopeptide-pulsed host myeloid and thymic dendritic cells. Diabetes 2001;50:1546–1552.

56 Turner T, Hackstein H, Wang Z, Thomson AW: Single infusion of pharmacologically-modified, alloantigen pulsed recipient DC prolongs transplant survival by inducing Ag-specific T cell hyporesponsiveness (Abstract). Am J Transpl 2004;4(suppl):#1611.

57 Mazariegos GV, Zahorchak AF, Reyes J, Ostrowski L, Flynn B, Zeevi A, Thomson AW: Dendritic cell subset ratio in peripheral blood correlates with successful withdrawal of immunosuppression in liver transplant patients. Am J Transplant 2003;3:689–696.

58 Billing J, Fry JW, Wheeler PR, Morris PJ, Wood KJ: Donor-specific dendritic cells modulate cardiac allograft rejection (abstract). Immunol 2001;104:L28.

59 Wang Z, Morelli AE, Hackstein H, Kaneko K, Thomson AW: Marked inhibition of transplant vascular sclerosis by in vivo-mobilised donor dendritic cells and anti-CD154 mAb. Transplantation 2003;76:562–571.

60 Thomas JM, Contreras JL, Jiang XL, et al: Peritransplant tolerance induction in macaque: Early events reflecting the unique synergy between immunotoxin and deoxyspergualin. Transplantation 1999;68:1660–1673.

61 Gorczynski RM, et al: Synergy in induction of increased renal allograft survival after portal vein infusion of dendritic cells transduced to express TGFb and IL-10, along with administration of CHO cells expressing the regulatory molecule OX-2. Clin Immunol 2000;95:182–189.

62 Mirenda V, Berton I, Read J, et al: Modified dendritic cells coexpressing self and allogeneic major histocompatability complex molecules: An efficient way to induce indirect pathway regulation. J Am Soc Nephrol 2004;15:987–997.

63 Coates PT, Duncan FJ, Colvin BL, Wang Z, Zahorchak AF, Shufesky WJ, Morelli AE, Thomson AW: In vivo-mobilized kidney dendritic cells are functionally immature, subvert alloreactive T-cell responses, and prolong organ allograft survival. Transplantation 2004;15;77:1080–1089.

64 Amsen D, Blander JM, Lee GR, Tanigaki K, Honjo T, Flavell RA: Instruction of distinct CD4 T helper cell fates by different notch ligands on antigen-presenting cells. Cell 2004;14;117:515–526.

65 Chen L: Co-inhibitory molecules of the B7-CD28 family in the control of T-cell immunity. Nat Rev Immunol 2004;4:336–347.

66 Demirci G, Amanullah F, Kewalaramani R, Yagita H, Strom TB, Sayegh MH, Li XC: Critical role of OX40 in CD28 and CD154-independent rejection. J Immunol 2004;172:1691–1698.

67 Gao W, Demirci G, Strom TB, Li XC: Stimulating PD-1-negative signals concurrent with blocking CD154 co-stimulation induces long-term islet allograft survival. Transplantation 2003;76:994–999.

Dr. Angus W. Thomson
W1540 Biomedical Science Tower
200 Lothrop St, Pittsburgh, PA 15213 (USA)
Tel. +1 412 624 6392, Fax +1 412 624 1172, E-Mail thomsonaw@upmc.edu

Ronco C, Chiaramonte S, Remuzzi G (eds): Kidney Transplantation: Strategies to Prevent
Organ Rejection. Contrib Nephrol. Basel, Karger, 2005, vol 146, pp 121–131

....

Natural versus Adaptive Regulatory T Cells

Linda Cassis, Sistiana Aiello, Marina Noris

'Center for Research on Organ Transplantation', Chiara Cucchi de Alessandri e
Gilberto Crespi, Mario Negri Institute for Pharmacological Research,
Villa Camozzi, Ranica (Bergamo), Italy

Abstract

It is now well recognized that regulatory T cells (Treg) play a central role in the control
of both reactivity to self-antigens and alloimmune response. Several subsets of Treg with
distinct phenotypes and mechanisms of action have now been identified. They constitute a
network of heterogeneous CD4+ or CD8+ T cell subsets and other minor T cell populations
such as nonpolymorphic CD1d-responsive natural killer T cells. Treg not only play a main
role in maintaining self-tolerance and preventing autoimmune disease but can also be
induced by tolerance protocols and seemed to play a key role in preventing allograft rejection, as demonstrated in many animal models. Of particular interest, in stable transplant
patients, CD4+CD25+ and CD8+CD28− Treg have been recently shown to modulate
immune response toward donor antigens in the indirect and direct pathway, respectively. This
finding raises the possibility that such Treg also have a role in the induction or maintenance
of transplant tolerance in humans.

Regulatory T Cells: Different Phenotypes and Mechanisms of Action

Suppression by regulatory T cells (Treg) has emerged as an essential tool
by which the immune system can actively either silence self-reactive T cells or
turn off activated T cells, thus controlling immune responses to self-antigens
and maintaining immune homeostasis. Several subsets of Treg with distinct
phenotypes and mechanisms of action have now been identified. They constitute a network of heterogeneous CD4+ [1–4] or CD8+ [5, 6] T cell subsets and
other minor T cell populations such as nonpolymorphic CD1d-responsive

Table 1. Natural regulatory T cells

Phenotype	Disease*(ref)	Suppressive mechanism	Regulatory factors
CD4+CD25+	Autoimmunity [10]	Cell contact-dependent	IL-10 TGF-β CTLA-4
NK T	Murine diabetes [20, 21] Murine GVHD [7]	Th2 cell-polarization	IFN-γ IL−4

GVHD: Graft-versus-host disease. * Disease associated with natural regulatory T cell deficit

natural killer T (NK T) cells [7, 8]. Treg can be distinguished into innate and adaptive. Innate Treg spontaneously arise during thymic ontogeny and are generated as a result of high-affinity interactions with cognate self-peptide/major histocompatibility complex (MHC) complexes within the thymus. In contrast, adaptive Treg have been shown to be specific for antigens not present in the thymus and similar to Th1 and Th2 cells, arise from naïve precursors and can be differentiated in vitro and in vivo.

Natural Treg (Table 1)

In both humans and rodents the best-characterized population of Treg are the CD4+CD25+ T cells, a subset of Treg constitutively coexpressing CD4 and CD25 (IL-2R α chain) antigens. CD4+CD25+ Treg, which constitute 5–10% of peripheral CD4+ T cells, are defined as 'naturally occurring' or 'innate' regulatory cells since they arise during thymic ontogeny, and are selected as a result of relatively high-affinity interactions with self-peptide/MHC complexes [9]. These cells play a main role in maintaining self-tolerance and preventing autoimmune diseases [2]. In naive mice, elimination of CD4+CD25+ Treg, by a thymectomy carried out at day 3 of age, induced the onset of a polyautoimmune syndrome [10]. Importantly, adoptive transfer of CD4+CD25+ T cells from naive mice to thymectomized animals protected from autoimmunity [10].

CD4+CD25+ Treg are anergic cells that, once activated, are able to inhibit both proliferation and cytokine production by CD4+ and CD8+ T cells in a cell contact-dependent and partially cytokine-independent manner. The contribution of cytokines, in particular TGF-β1, to the suppressive activity is a controversial issue. Murine CD4+CD25+ Treg are characterized by high levels of TGF-β1, in a cell surface bound form, and the ability of these cells to suppress CD25− T cell proliferation is abolished by an anti-TGF-β1 antibody [11, 12]. On the other hand CD4+CD25+ Treg from TGF-β1-deficient mice can suppress naïve

CD25− T cells as well [13], which would suggest that TGF-β is dispensable to regulatory activity.

The main mechanism of suppression by CD4+CD25+ Treg seems to be the inhibition of IL-2 production by responder T cells [14]. Interestingly, both in mice and in humans CD4+CD25+ Treg have been shown to constitutively express CTLA4 (CD152). Recently, Fallarino et al. [15] demonstrated that mouse CD4+CD25+ Treg block the immunostimulatory function of antigen presenting cells (APCs) through CTLA4 engagement of B7 molecule which attributes a key role to CTLA4 in Treg function. Thus, CD4+CD25+ Treg can exert their regulatory activity either by directly suppressing T cells or indirectly through modulation of APC function.

Identification of a specific marker for Treg remains a controversial issue since activated effector CD4+ T cells also express CD25. Finding that mice carrying the X-linked *scurfy* mutation in FoxP3 gene display multiorgan autoimmune disease and lack conventional CD4+CD25+ Treg [16, 17] has focused the attention on FoxP3 as a specific marker of Treg in mice. In mice, FoxP3 has been shown to be expressed exclusively by CD4+CD25+ Treg and is not induced upon activation of CD25− T cells. In addition, transfection with FoxP3 converts naïve CD4+CD25− T cells into Treg [18]. Of particular interest, Walker et al. [19] have recently shown that in humans, activation of CD4+CD25− T cells results in the raise of two populations of cells, effector CD4+CD25+ and regulatory CD4+CD25+ T cells, with expression of FoxP3 confined to the regulatory cell subpopulation.

A second type of spontaneously arising Treg has emerged in recent years that is part of the innate immune system, the NK T cells. Indeed, finding that autoimmune-prone mouse strains, such as nonobese diabetic mice, have a numerical and functional NK T cell deficiency, have proposed NK T cells as another 'naturally occurring' Treg subset [20]. NK T cells recognize CD1d, a nonpolymorphic class I MHC-like antigen-presenting molecule that binds glycolipids. Activation of NK T cells, followed by the release of large quantities of IFN-γ and IL-4, has been shown to ameliorate autoimmune diabetes in nonobese diabetic mice by polarizing immune responses to a T helper type 2 pattern [21].

In a model of graft-versus-host disease (GVHD) induced by infusion of bone marrow T cells from normal mice (C57BL/6) in lethally irradiated hosts (BALB/c) [7] the depletion of NK T cells from bone marrow before infusion increased the percentage of hosts who died by GVHD. Add-back of bone marrow-derived NK T cells protected hosts against lethal GVHD. However, add-back of NK T cells from IL-4$^{-/-}$ donors failed to provide protection, indicating that IL-4 plays a key role in the inhibition of lethal GVHD by NK T cells.

Table 2. Adaptive regulatory T cells

Phenotype	Disease/Experimental model (ref)	Differentiation factors	Suppressive mechanism	Regulatory factors
Th3	Murine EAE [22] Human multiple sclerosis [23]	TGF-β IL-4?	Cell contact-independent TGF-β-dependent	TGF-β
Trl	Murine colitis [24] Murine EAE [27]	IL-10 Dex+Vit D3 CD3/CD46 stimulation IL-10 + IFN-α Immature DCs	Cell contact-independent	IL-10 TGF-β
CD8+	?	CD40L-activated DC2 Injection of immature DCs	Cell contact-independent	IL-10
CD8+CD28−	Human transplantation [46]	?	Cell contact-dependent	ILT3 ILT4
CD4+CD25+	Transplantation [36, 37, 40, 41] Murine GVHD [42]	?	Cell contact-dependent	IL-10 TGF-β CTLA-4

EAE: Experimental autoimmune encephalomyelitis; GVHD: graft-versus-host disease; Dex: dexamethasone; DC: dendritic cell; ILT: immunoglobulin-like transcript.

Adaptive Treg (Table 2)

In addition to naturally occurring Treg, it appears to be possible to steer an uncommitted T cell toward regulatory function (induced or adaptive Treg). Adaptive Treg can be generated either in vivo, from mature CD4+ T cell populations under particular conditions of antigenic stimulations, or ex vivo by culturing naive CD4+ T cells with antigen or polyclonal activators in the presence of immunosuppressive factors. Antigen exposure by oral administration has been shown to induce selectively the appearance of CD4+ T cells with regulatory properties. These Treg, named Th3 cells, were originally generated and identified in mice orally tolerized to myelin basic protein (MBP) [22]. After treatment with MBP, the majority of MBP-specific CD4+ T cells secrete TGF-β and suppress the induction of a MBP-specific experimental autoimmune encephalitis (EAE) in vivo [22]. This suppression is abrogated by injection of anti-TGF-β antibodies.

Furthermore, these Th3 cells suppress the proliferation and cytokine release of MBP-specific Th1 cells in vitro in a TGF-β-dependent manner [23].

Another CD4+ T cell subset with suppressive activity has been induced in vitro by antigenic stimulation of naïve CD4+ T cells in the presence of IL-10 [24, 25]. These Treg, designed T regulatory type 1 cells (Tr1), are characterized by a unique cytokine profile distinct from that of Th0, Th1, or Th2 cells. They produce IL-10, TGF-β, some IL-5 and IFN-γ, and little or no IL-2 and IL-4 [3, 26], express very low levels of CD25 in resting conditions [25], do not proliferate in response to IL-2 unless at extremely high concentrations [25] and strongly suppress the activity of both Th1 and Th2 T cells through the release of IL-10 and TGF-β in a completely cell-contact-independent manner.

Tr1 cells have been shown to prevent the development of colitis induced by the transfer of naïve CD45RB[hi] cells into SCID mice, a model of Th1-mediated autoimmune disease [24]. In addition, Tr1 cells differentiated with dexamethasone and vitamin D3 suppressed the induction of EAE in mice [27]. Protection from EAE was dependent on the presence of the antigen being recognized by Tr1 cells, indicating that they must be activated via TCR in order to exert their regulatory effect. However, once activated Tr1 cells suppress T cell response in an antigen-nonspecific manner as documented by data that Tr1 clones specific for filamentous hemagglutinin from *Bordetella pertussis*, inhibited proliferation and cytokine production by a Th1 clone against an unrelated antigen, influenza virus hemagglutinin [28]. Recently Tr1 cells have also been obtained from human CD4+ T cells by coengagement of CD3 and the complement regulator CD46 in the presence of IL-2 [29] suggesting a role for CD46 in human T cell regulation and establishing a link between the complement system and adaptive immunity.

It has been proposed that Tr1 cells derive from CD4+CD25+ Treg that emerge from the thymus in a partially differentiated state and terminally mature into IL-10 and TGF-β-producing Tr1 cells only upon encountering antigens in the periphery [30]. This possibility has been refuted by recent data showing that human CD4+CD25− cells can be differentiated in vitro into Tr1 cells by IL-10 and IFN-α in the absence of CD4+CD25+ T cells [31]. Whether Tr1 cell generation could be induced by a dedicated cell population has also been investigated. Recent data suggest that immature dendritic cells (DCs), i.e. under steady state conditions, play a crucial role in maintaining self-tolerance by inducing the formation of Tr1 cells. Indeed, Jonuleit et al. [32] showed that repetitive stimulation of naïve cord blood-derived CD4+ T cells by allogeneic immature dendritic cells generated IL-10 producing T cells displaying most of the typical properties of Tr1 cells. Recent evidence document that human CD4+CD25+ Treg, besides inhibiting the proliferative response of naïve CD4+ T cells, modify their phenotype and generate IL-10-producing Tr1-like cells [33, 34]. The above data create a functional link between the two Treg populations suggesting that tolerance to peripheral antigens is a well-orchestrated process.

Beside naturally occurring, CD4+CD25+ Treg can also be induced by tolerance protocols and play a role in preventing allograft rejection, as demonstrated in many animal models [35]. In this regard, in a model of rat kidney allograft tolerance induced by preinfusion of donor peripheral blood leukocytes, it has been shown that lymph node cells from long-term surviving rats inhibit naïve T cell proliferation against donor antigens, and that this immunoregulatory activity is confined to the CD4+CD25+ subset [36]. Furthermore, it has been demonstrated that CD4+CD25+ Treg with the capacity to prevent skin allograft rejection can be generated in mice by pretreatment with donor alloantigen under the cover of nondepleting anti-CD4 therapy [37]. CD4+CD25+ Treg isolated from the spleens of these tolerant mice are donor specific and can transfer tolerance when infused into a naïve recipient [38]. Of great interest, the same group has recently shown that such Treg are generated in the periphery from CD4+CD25− precursors indicating that their ontogeny is distinct from that of naturally occurring CD4+CD25+ Treg [39].

Interestingly, Cobbold et al. [40] recently investigated whether CD4+CD25+ Treg induced by a nondepleting anti-CD4 mAb tolerance protocol express FoxP3 like their naturally occurring counterpart. The authors used a model of skin graft with female transgenic mice, which have no detectable pre-existing CD4+CD25+FoxP3+ Treg in the thymus or periphery, as recipients. Long-term skin graft tolerance was associated with the presence within the graft of Treg that expressed CD4, CD25 and high levels of FoxP3 mRNA, and that would appear to have arisen de novo in the periphery. In a previous study, however, the role of pre-existing natural CD4+CD25+ Treg in generating allograft tolerance under the cover of CD4-targeted therapy has been proposed. Thymectomy before, but not after, transplantation prevents the induction and generation of Treg in CD4-mAb-treated rat recipients, suggesting that this Treg are derived from recent thymus emigrants [41].

Induced CD4+CD25+ Treg have been shown to play an important role also in allogeneic bone marrow transplantation by protecting recipients from acute GVHD. In this setting, emerging data demonstrate that infusion of ex vivo-activated and expanded donor CD4+CD25+ Treg can offer substantial protection in an in vivo model of the disease in mice [42].

Immunoregulatory activity is not exclusively confined into CD4+ T cells; indeed data on the existence of a subset of CD8+ T cells with strong regulatory properties have now been emerging. In humans, IL-10 producing CD8+ Treg have been induced either in vitro by interaction of naïve CD8+ T cells with CD40-L-activated plasmacytoid dendritic cells [6] or in vivo by injection of immature dendritic cells into healthy volunteers [43]. Like Tr1 cells, these

CD8+ Treg exert their suppressive activity in a cell-contact-independent manner. Another subset of CD8+ Treg has been found recently. They are characterized by the lack of CD28 receptor and are referred to as CD8+CD28− Treg [44]. CD8+CD28− Treg suppressive activity is cell-contact dependent; they recognize MHC class I peptide complexes on APCs, rendering them tolerogenic by up-regulation of inhibitory receptors such as immunoglobulin-like transcripts 3 and 4 (ILT3 and ILT4). In turn, ILT3 and ILT4 overexpression prevents APC up-regulation of costimulatory molecules, such as CD80, induced by allogeneic CD4+ T cells [45, 46].

Regulatory Cells in Transplant Patients

A large fraction of stable transplant patients show a low in vitro alloreactivity [47] that in the past has been attributed to T cell anergy [48]. However, as suggested by animal models [38], a low reactivity toward donor antigens may also reflect the presence of Treg.

Efforts to study the role and the relevance of CD4+CD25+ Treg in the regulation of alloimmune responses in transplant patients has only recently emerged. The frequency and functional profile of circulating CD4+CD25+ T cells have been evaluated in 10 lung transplant recipients with stable clinical condition and in 11 patients with clinical signs of chronic rejection. The frequency of CD4+CD25+ T cells was significantly higher in stable transplant patients as compared with that recorded in patients with chronic rejection. In addition, functional evaluation of these cells demonstrated their regulatory profile: they were hyporesponsive to conventional T cell stimuli and suppressed the proliferation of CD4+CD25− T cells [49].

To better clarify the function of CD4+CD25+ Treg in clinical transplantation, other authors investigated their role in regulating either the direct or the indirect pathway of alloimmune activation. The effect of Treg on the direct pathway was evaluated on peripheral blood leukocytes isolated from 12 stable renal transplant patients by using mixed leukocyte culture, limiting dilution assay, and ELISPOT for INF-γ. Depletion of CD4+CD25+ cells from patients' peripheral blood leukocytes did not increase the low frequency of donor-specific alloreactive T cell clones, thus excluding a role of CD4+CD25+ Treg in maintaining hyporesponsiveness [50]. On the other hand, other authors have suggested that CD4+CD25+ Treg may control T cell response through the indirect pathway. In stable renal transplant patients, chosen for having low reactivity to the mismatched donor-derived HLA-DR antigens, Salama et al. [51] detected significant increase in the frequency of IFN-γ-producing T cells after depletion of the CD25+ subset.

As mentioned above, CD8+CD28− Treg display regulatory function by inducing up-regulation of inhibitory receptors such as ILT3 and ILT4 on APCs [45, 52]. Interestingly, in humans, up-regulation of ILT3 and ILT4 on donor monocytes has been correlated with the absence of acute rejection in heart allograft recipients [46], which suggest a role of CD8+CD28− Treg in maintaining hyporesponsiveness.

It has been shown that human CD8+CD28− Treg arise in the course of repeated in vitro allostimulations which lead to hypothesize that they may also develop in vivo in recipients of allogeneic transplants. In this regard, Ciubotariu et al. [53], by performing flow cytometry analysis of blood samples from heart, liver, and kidney transplant recipients, detected donor-specific CD8+CD28− Treg in all patients with a stable graft function. In contrast, these cells were not detectable in the circulation of patients undergoing acute rejection. These data provided the evidence that the presence of CD8+CD28− is relevant to the outcome of transplants and that these cells participate in the induction and maintenance of peripheral tolerance.

CD8+CD28− Treg act by inhibiting the activity of donor-derived APCs and block the direct pathway, thus their activity could be considered as complementary to that of CD4+CD25+ cells, which have been demonstrated to act mainly in the control of the indirect pathway.

Acknowledgements

This paper has been partially supported by Transplant Research Association (ART, Milan, Italy) and by 'Fondazione Italo Monzino' (Milan, Italy).

References

1   Shevach EM: Certified professionals: CD4$^+$CD25$^+$ suppressor T cells. J Exp Med 2001;193: F41–46.
2   Sakaguchi S: Regulatory T cells: Key controllers of immunologic self-tolerance. Cell 2000; 101:455–458.
3   Roncarolo MG, Bacchetta R, Bordignon C, Narula S, Levings MK: Type 1 T regulatory cells. Immunol Rev 2001;182:68–79.
4   Weiner HL: Induction and mechanism of action of transforming growth factor-beta-secreting Th3 regulatory cells. Immunol Rev 2001;182:207–214.
5   Ciubotariu R, Colovai AI, Pennesi G, Liu Z, Smith D, Berlocco P, Cortesini R, Suciu-Foca N: Specific suppression of human CD4$^+$ Th cell responses to pig MHC antigens by CD8$^+$CD28$^-$ regulatory T cells. J Immunol 1998;161:5193–5202.
6   Gilliet M, Liu YJ: Generation of human CD8 T regulatory cells by CD40 ligand-activated plasmacytoid dendritic cells. J Exp Med 2002;195:695–704.
7   Zeng D, Lewis D, Dejbakhsh-Jones S, Lan F, Garcia-Ojeda M, Sibley R, Strober S: Bone marrow NK1.1$^-$ and NK1.1$^+$ T cells reciprocally regulate acute graft versus host disease. J Exp Med 1999;189:1073–1081.

8   Seino KI, Fukao K, Muramoto K, Yanagisawa K, Takada Y, Kakuta S, Iwakura Y, Van Kaer L, Takeda K, Nakayama T, et al: Requirement for natural killer T (NK T) cells in the induction of allograft tolerance. Proc Natl Acad Sci USA 2001;98:2577–2581.

9   Jordan MS, Boesteanu A, Reed AJ, Petrone AL, Holenbeck AE, Lerman MA, Naji A, Caton AJ: Thymic selection of CD4⁺CD25⁺ regulatory T cells induced by an agonist self-peptide. Nat Rev Immunol 2001;2:301–306.

10 Sakaguchi S, Sakaguchi N, Asano M, Itoh M, Toda M: Immunologic self-tolerance maintained by activated T cells expressing IL-2 receptor alpha-chains (CD25). Breakdown of a single mechanism of self-tolerance causes various autoimmune diseases. J Immunol 1995;155:1151–1164.

11 Nakamura K, Kitani A, Strober W: Cell contact-dependent immunosuppression by CD4⁺CD25⁺ regulatory T cells is mediated by cell surface-bound transforming growth factor beta. J Exp Med 2001;194:629–644.

12 Nakamura K, Kitani A, Fuss I, Pedersen A, Harada N, Nawata H, Strober W: TGF-beta1 plays an important role in the mechanism of CD4⁺CD25⁺ regulatory T cell activity in both humans and mice. J Immunol 2004;172:834–842.

13 Piccirillo CA, Letterio JJ, Thornton AM, McHugh RS, Mamura M, Mizuhara H, Shevach EM: CD4⁺CD25⁺ regulatory T cells can mediate suppressor function in the absence of transforming growth factor beta1 production and responsiveness. J Exp Med 2002;196:237–246.

14 Thornton AM, Shevach EM: Suppressor effector function of CD4⁺CD25⁺ immunoregulatory T cells is antigen nonspecific. J Immunol 2000;164:183–190.

15 Fallarino F, Grohmann U, Hwang KW, Orabona C, Vacca C, Bianchi R, Belladonna ML, Fioretti MC, Alegre ML, Puccetti P: Modulation of tryptophan catabolism by regulatory T cells. Nat Immunol 2003;4:1206–1212.

16 Khattri R, Cox T, Yasayko SA, Ramsdell F: An essential role for Scurfin in CD4⁺CD25⁺ T regulatory cells. Nat Rev Immunol 2003;4:337–342.

17 Fontenot JD, Gavin MA, Rudensky AY: Foxp3 programs the development and function of CD4⁺CD25⁺ regulatory T cells. Nat Immunol 2003;4:330–336.

18 Hori S, Nomura T, Sakaguchi S: Control of regulatory T cell development by the transcription factor Foxp3. Science 2003;299:1057–1061.

19 Walker MR, Kasprowicz DJ, Gersuk VH, Benard A, Van Landeghen M, Buckner JH, Ziegler SF: Induction of FoxP3 and acquisition of T regulatory activity by stimulated human CD4⁺CD25⁻ T cells. J Clin Invest 2003;112:1437–1443.

20 Gombert JM, Herbelin A, Tancrede-Bohin E, Dy M, Carnaud C, Bach JF: Early quantitative and functional deficiency of NK1⁺-like thymocytes in the NOD mouse. Eur J Immunol 1996;26: 2989–2998.

21 Sharif S, Arreaza GA, Zucker P, Mi QS, Sondhi J, Naidenko OV, Kronenberg M, Koezuka Y, Delovitch TL, Gombert JM, et al: Activation of natural killer T cells by alpha-galactosylceramide treatment prevents the onset and recurrence of autoimmune Type 1 diabetes. Nat Med 2001;7: 1057–1062.

22 Chen Y, Kuchroo VK, Inobe J, Hafler DA, Weiner HL: Regulatory T cell clones induced by oral tolerance: Suppression of autoimmune encephalomyelitis. Science 1994;265:1237–1240.

23 Fukaura H, Kent SC, Pietrusewicz MJ, Khoury SJ, Weiner HL, Hafler DA: Induction of circulating myelin basic protein and proteolipid protein-specific transforming growth factor-beta1-secreting Th3 T cells by oral administration of myelin in multiple sclerosis patients. J Clin Invest 1996;98: 70–77.

24 Groux H, O'Garra A, Bigler M, Rouleau M, Antonenko S, de Vries JE, Roncarolo MG: A CD4⁺ T-cell subset inhibits antigen-specific T-cell responses and prevents colitis. Nature 1997;389: 737–742.

25 Bacchetta R, Sartirana C, Levings MK, Bordignon C, Narula S, Roncarolo MG: Growth and expansion of human T regulatory type 1 cells are independent from TCR activation but require exogenous cytokines. Eur J Immunol 2002;32:2237–2245.

26 Groux H: Type 1 T-regulatory cells: Their role in the control of immune responses. Transplantation 2003;75:S8–S12.

27 Barrat FJ, Cua DJ, Boonstra A, Richards DF, Crain C, Savelkoul HF, de Waal-Malefyt R, Coffman RL, Hawrylowicz CM, O'Garra A: In vitro generation of interleukin 10-producing regulatory

CD4$^+$ T cells is induced by immunosuppressive drugs and inhibited by T helper type 1 (Th1)- and Th2-inducing cytokines. J Exp Med 2002;195:603–616.

28 McGuirk P, McCann C, Mills KH: Pathogen-specific T regulatory 1 cells induced in the respiratory tract by a bacterial molecule that stimulates interleukin 10 production by dendritic cells: A novel strategy for evasion of protective T helper type 1 responses by Bordetella pertussis. J Exp Med 2002;195:221–231.

29 Kemper C, Chan AC, Green JM, Brett KA, Murphy KM, Atkinson JP: Activation of human CD4$^+$ cells with CD3 and CD46 induces a T-regulatory cell 1 phenotype. Nature 2003;421:388–392.

30 Roncarolo MG, Levings MK, Traversari C: Differentiation of T regulatory cells by immature dendritic cells. J Exp Med 2001;193:F5–F9.

31 Levings MK, Sangregorio R, Sartirana C, Moschin AL, Battaglia M, Orban PC, Roncarolo MG: Human CD25$^+$CD4$^+$ T suppressor cell clones produce transforming growth factor beta, but not interleukin 10, and are distinct from type 1 T regulatory cells. J Exp Med 2002;196: 1335–1346.

32 Jonuleit H, Schmitt E, Schuler G, Knop J, Enk AH: Induction of interleukin 10-producing, non-proliferating CD4$^+$ T cells with regulatory properties by repetitive stimulation with allogeneic immature human dendritic cells. J Exp Med 2000;192:1213–1222.

33 Stassen M, Schmitt E, Jonuleit H: Human CD4$^+$CD25$^+$ regulatory T cells and infectious tolerance. Transplantation 2004;77:S23–S25.

34 Dieckmann D, Bruett CH, Ploettner H, Lutz MB, Schuler G: Human CD4$^+$CD25$^+$ regulatory, contact-dependent T cells induce interleukin 10-producing, contact-independent type 1-like regulatory T cells. J Exp Med 2002;196:247–253.

35 Waldmann H, Cobbold S: Regulating the immune response to transplants. A role for CD4+ regulatory cells? Immunity 2001;14:399–406.

36 Cavinato RA, Casiraghi F, Noris M, Azzollini N, Aiello S, Mister M, Pezzotta A, Cugini D, Remuzzi G: Tolerance induction by peripheral mononuclear leukocyte (PBMC) infusion is associated with the generation of immunoregulatory T cells. Transplantation 2002;74:156.

37 Kingsley CI, Karim M, Bushell AR, Wood KJ: CD25$^+$CD4$^+$ regulatory T cells prevent graft rejection: CTLA-4- and IL-10-dependent immunoregulation of alloresponses. J Immunol 2002; 168:1080–1086.

38 Wood KJ, Sakaguchi S: Regulatory T cells in transplantation tolerance. Nat Rev Immunol 2003; 3:199–210.

39 Karim M, Kingsley CI, Bushell AR, Sawitzki BS, Wood KJ: Alloantigen-induced CD25+CD4+ regulatory T cells can develop in vivo from CD25$^-$CD4$^+$ precursors in a thymus-independent process. J Immunol 2004;172:923–928.

40 Cobbold SP, Castejon R, Adams E, Zelenika D, Graca L, Humm S, Waldmann H: Induction of foxP3$^+$ regulatory T cells in the periphery of T cell receptor transgenic mice tolerized to transplants. J Immunol 2004;172:6003–6010.

41 Onodera K, Volk HD, Ritter T, Kupiec-Weglinski JW: Thymus requirement and antigen dependency in the 'infectious' tolerance pathway in transplant recipients. J Immunol 1998;160: 5765–5772.

42 Taylor PA, Lees CJ, Blazar BR: The infusion of ex vivo activated and expanded CD4$^+$CD25$^+$ immune regulatory cells inhibits graft-versus-host disease lethality. Blood 2002;99:3493–3499.

43 Dhodapkar MV, Steinman RM, Krasovsky J, Munz C, Bhardwaj N: Antigen-specific inhibition of effector T cell function in humans after injection of immature dendritic cells. J Exp Med 2001; 193:233–238.

44 Liu Z, Tugulea S, Cortesini R, Suciu-Foca N: Specific suppression of T helper alloreactivity by allo-MHC class I-restricted CD8$^+$CD28$^-$ T cells. Int Immunol 1998;10:775–783.

45 Chang CC, Ciubotariu R, Manavalan JS, Yuan J, Colovai AI, Piazza F, Lederman S, Colonna M, Cortesini R, Dalla-Favera R, et al: Tolerization of dendritic cells by T(S) cells: The crucial role of inhibitory receptors ILT3 and ILT4. Nat Rev Immunol 2002;3:237–243.

46 Colovai AI, Mirza M, Vlad G, Wang S, Ho E, Cortesini R, Suciu-Foca N: Regulatory CD8$^+$CD28$^-$ T cells in heart transplant recipients. Hum Immunol 2003;64:31–37.

47 Mason PD, Robinson CM, Lechler RI: Detection of donor-specific hyporesponsiveness following late failure of human renal allografts. Kidney Int 1996;50:1019–1025.

48   Ng WF, Hernandez-Fuentes M, Baker R, Chaudhry A, Lechler RI: Reversibility with interleukin-2 suggests that T cell anergy contributes to donor-specific hyporesponsiveness in renal transplant patients. J Am Soc Nephrol 2002;13:2983–2989.

49   Meloni F, Vitulo P, Bianco AM, Paschetto E, Morosini M, Cascina A, Mazzucchelli I, Ciardelli L, Oggionni T, Fietta AM, et al: Regulatory CD4+CD25+ T cells in the peripheral blood of lung transplant recipients: Correlation with transplant outcome. Transplantation 2004;77:762–766.

50   Game DS, Hernandez-Fuentes MP, Chaudhry AN, Lechler RI: CD4+CD25+ regulatory T cells do not significantly contribute to direct pathway hyporesponsiveness in stable renal transplant patients. J Am Soc Nephrol 2003;14:1652–1661.

51   Salama AD, Najafian N, Clarkson MR, Harmon WE, Sayegh MH: Regulatory CD25+ T cells in human kidney transplant recipients. J Am Soc Nephrol 2003;14:1643–1651.

52   Cortesini R, LeMaoult J, Ciubotariu R, Cortesini NS: CD8+CD28– T suppressor cells and the induction of antigen-specific, antigen-presenting cell-mediated suppression of Th reactivity. Immunol Rev 2001;182:201–206.

53   Ciubotariu R, Vasilescu R, Ho E, Cinti P, Cancedda C, Poli L, Late M, Liu Z, Berloco P, Cortesini R, et al: Detection of T suppressor cells in patients with organ allografts. Hum Immunol 2001;62:15–20.

Dr. Marina Noris
Mario Negri Institute for Pharmacological Research
Villa Camozzi, via Camozzi 3
IT–24024 Ranica, Bergamo (Italy)
Tel. +39 035 4535 362, Fax +39 035 4535 377, E-Mail noris@marionegri.it

Ronco C, Chiaramonte S, Remuzzi G (eds): Kidney Transplantation: Strategies to Prevent
Organ Rejection. Contrib Nephrol. Basel, Karger, 2005, vol 146, pp 132–142

....

Reviewing the Mechanism of Peripheral Tolerance in Clinical Transplantation

Nicole Suciu-Foca, Raffaello Cortesini

Columbia University, Department of Pathology, New York, N.Y., USA

Introduction

The immune system has developed several checkpoints and regulatory systems to discriminate between self and nonself antigens (Ags) and avoid autoimmunity. Included among them are: (1) negative selection of autoreactive T cells in the thymus [1]; (2) elimination of autoreactive T cells via activation-induced cell death [2–6]; (3) induction of anergy upon TCR triggering in the absence of costimulation [7–9]; (4) inhibition of immune function by T suppressor cells [10].

Research on T suppressor cells has re-emerged in the late 1990s when several subsets of T cells were shown to inhibit the proliferation of other cells. Two broad categories of regulatory T cells (T regs) have been recognized. The first consists of naturally occurring and the second, of induced T regs [10–12].

The naturally occurring CD4+CD25+ T reg subset is generated in the thymus as a functionally distinct subpopulation of T cells [10–18]. These natural T regs play a major role in regulating self-reactive T cells and preventing autoimmune diseases. The transfer of T cell populations, from which the CD25+ subset has been depleted, into T cell-deficient mice caused severe autoimmune disorders, such as thyroiditis, gastritis, insulin-dependent diabetes mellitus and colitis [13, 16, 17]. Conversely, infusion of the T reg cells in these animals strongly suppressed autoimmunity. Recent evidence indicates that the transcription factor FOXP3 acts as the 'master control gene' for T regs [18–20]. A mutation in the gene encoding FOXP3 was identified as the genetic defect underlying autoimmune and inflammatory disease in scurvy mice and in humans with immune dysregulation, polyendocrinopathy, enteropathy X-linked syndrome, IPEX, and X-linked autoimmunity allergic dysregulation syndrome,

(XLAAD) emphasizing the importance of CD4+CD25+ T regs in the maintenance of normal immune homeostasis.

FOXP3 is a member of the forkhead–winged helix family of transcription factors. While in mice FOXP3 is stably expressed by CD4+CD25+ T regs and cannot be induced by activation of CD4+CD25− T cells, in humans it can be induced by activation and exposure to transforming growth factor-β (TGF-β).

CD4+CD25+ T regs suppress activation and proliferation of naive CD4+ T cells in vitro through cell-cell contact [18]. T reg cells, because they are thought to be in an anergic state, do not proliferate in response to antigens (Ags) presented by MHC class II molecules [18]. Triggering of T reg cells by anti-CD3 plus anti-CD28 monoclonal antibodies or exposure to IL-6 breaks the anergic state and abrogates the suppressor function of T reg [19, 20]. Also, triggering by an agonistic antibody of the cell surface receptor glucocorticoid-induced TNFR-related protein (GITR; TNFRSF18), which is constitutively expressed on the surface of T reg cells [21, 22] abrogates the suppressor activity of T regs. The mechanisms by which natural CD4+CD25+ T regs act in different autoimmune disease models seem to involve a discretionary requirement for the cytokines IL-10 and TGF-β [14], and may depend on signals through CTLA-4 [19, 23, 24] as well as regulation coordinated by GITR [25, 26].

Induced T reg cells fall into two categories, one consisting of non-Ag-specific CD4+ T reg and the other of Ag-specific CD8+CD28− T suppressor cells (T_S) and CD4+CD25+ T regs.

Non-Ag-specific regulatory CD4+CD25+ T cells can be generated from naïve peripheral CD25− precursors under certain culture conditions, such as by exposure to IL-10 and interferon-α (IFN-α) or TGF-β, or by stimulation with immature myeloid dendritic cells (DC) or mature plasmacytoid DC [10]. Activation, in vitro or in vivo, of human or mouse CD4+ T cells in the presence of IL-10 results in the generation of T cell clones which produce significant amounts of IL-10, IFN-α, TFG-β and IL-5. These T cell clones, named T_{R1}, inhibit Ag-induced activation of naïve autologous T cells via a mechanism, which is partially mediated by IL-10 and TGF-β [10, 27]. T cells with a T_{R1} cytokine profile have been described in several models of autoimmune diseases and transplantation [28–32]. In most cases, T_{R1} cells arise following repeated Ag stimulation either in vitro or in vivo [27]. However, their inhibitory effect on other T cells is not Ag specific.

The extent to which cell contact or soluble factors are required for T_{R1}-mediated suppression of T_H reactivity is still unknown. Experiments in which cytokine production was excluded by first activating and then fixing naturally occurring CD4+CD25+ T_R cells showed that these cells maintain their capacity to suppress normally responding CD4+CD25− population rendering them anergic. The newly anergized population further suppressed syngeneic

CD4+ T cells via the production of inhibitory cytokines [33, 34]. It was suggested that suppression occurs in two sequential steps: The first one is a cell contact-dependent 'transmission' of anergy from a T reg to another T lymphocyte, while the second is a cell contact-independent, cytokine-mediated suppression of other T helper cells [33, 34].

Other non-Ag-specific T reg subtypes have been described including γδ, NK1.1 T cells, CD8+CD25+ and CD4−CD8− T cells, but they remain poorly characterized [11].

A distinct category of T regs characterized by their Ag-specific activity and mechanism of action is CD8+CD28− T_S cells first described by our group. We showed that CD8+CD28− T_S specific for alloantigens, xenoantigens or nominal Ags could be generated in vitro by repeated antigenic stimulation [35–46]. Next, we provided in vivo evidence that CD8+CD28− T cells act as suppressors in patients with heart, kidney or liver allografts [41, 43–48]. We further demonstrated that Ag-specific CD8+CD28− T_S express FOXP3$^+$, derived from an oligoclonal population of CD8+FOXP3− cells, are MHC class I restricted, have no killing capacity and do not produce cytokines [35–46]. Instead they act on professional (DC) and nonprofessional (endo-thelial cells) antigen presenting cells (APC) directly i.e., by cell-to-cell con-tact, inducing qualitative changes characteristic of an alternative pathway of maturation toward a tolerogenic rather than immunogenic phenotype. These changes include the down-regulation of NF-κB-dependant costimulatory mol-ecules such as CD40, CD58, CD80, CD86, and the up-regulation of inhibitory receptors immunoglobulin-like transcript (ILT)3 and ILT4 [41, 43–48].

CD4+ T_H, which interacts with tolerogenic APC, become anergic and acquires regulatory activity. Our data support a model in which T cell-mediated suppression results from the sequential interaction between first, T_S and APCs and next, 'tolerized' APCs and T_H. In turn, anergic T_H acquire regulatory capacity, in conjunction with FOXP3 expression, and further perpetuate tolerance [10, 47]. The central role of ILT3^{high}, ILT4^{high} APC in the induction of suppression was further demonstrated in experiments in which we induced the up-regulation of these inhibitory receptors in DC by treating the cells with IL-10 plus IFN-α or IL-10 plus vitamin D3. Such ILT3^{high} ILT4^{high} DC induced the in vitro generation of T_S and T_R from unprimed populations of CD4$^+$ and CD8$^+$ T cells [10, 39, 47, 48].

Progress in understanding the mechanisms of Tcell activation and inacti-vation is currently being translated into strategies, enabling induction of selec-tive immunosuppression for treatment of autoimmune diseases, allergies and allograft rejection. There is an imperative need for Ag-specific immunosup-pression, as systemic immunosuppression is associated with an increased risk of malignancies, infection and considerable toxicity. Progress in the generation and characterization of T_S, T regs and tolerogenic APC may pave the way to the induction of immunological tolerance.

Molecular Characterization of T_S

To gain insight into the common denominators of Ag-specific and nonspecific T reg cells, we have analyzed some of their characteristics at the molecular level. We compared the expression of genes, known to be up-regulated in natural CD4+CD25+ T reg cells from fresh peripheral blood [18, 23, 49–55] with their expression in allospecific CD8+CD28− T_S from T cell lines (TCL) [56]. The following genes were studied by RT-PCR: CD25, GITR, CTLA-4, FOXP3, CD62L, OX40, 4–1BB, TNFR2 and CD103. All of these genes showed similar levels of expression in natural T reg cells- and Ag-induced T_S. The exceptions were: CD25, which showed higher expression in natural compared to induced T reg cells; CD62L and 4–1BB which showed lower expression in CD8$^+$ compared to CD4+CD25+ T_R. Unprimed CD8+CD28− T cells from fresh peripheral blood, which have no regulatory function, do not express FOXP3, GITR, OX40, CD25, CD62L and 4–1BB. Compared to the regulatory cells they show low levels of CTLA-4 and TNFR2, yet similar levels of CD103 expression [56].

To examine the structure of the FOXP3 transcript expressed by CD8+CD28− T_S from allospecific TCL and T cell clones TCC, we used primers complementary to the 5′ and 3′ untranslated regions of the FOXP3 transcript and amplified the coding region of this gene. Sequence analysis of seven isolated clones and comparison with the GenBank database indicated the presence of a novel, alternatively spliced form of the previously described FOXP3 gene. Alignment of amino acid sequences of the two FOXP3 isoforms showed that the newly identified isoform lacks exon 3 encoding a 35-amino acid region corresponding to position 71–105 of the previously described FOXP3 protein product. This isoform is also expressed in natural CD4+CD25+ T_R. However, neither CD4+CD25− nor CD8+CD28− and CD8+CD28+ T cells from fresh peripheral blood of healthy adult individuals express either FOXP3 or FOXPα [48]. Study of CD8+CD28− T cell clones derived from allospecific TCL showed that in most T cell clones the two isoforms were coexpressed.

In a further attempt to better define CD8+CD28− T_S and identify genes which are important for their function we performed mRNA microanalysis of CD8+CD28− and CD8+CD28+ T cells from five different TCL [56].

Affymetrix gene chip analysis of 12,000 genes showed that 72 genes were differentially expressed in CD8+CD28− T cells compared to CD8+CD28+ T cells. Among the genes with higher expression in the CD8+CD28− T cell subset, three were members of the killing inhibitory receptors (KIR) family: KIR3DL1 (NKAT3, CD158E2), KIR3DL2 (NKAT4, CD158K) and KIR2DL3 (NKAT2, CD158B2) [56].

The oncogene LYN (V-YES-1 Yamaguchi Sarcoma Viral Related Oncogene Homolog) was also expressed at higher levels in CD8+CD28− T_S compared to CD8+CD28− cytotoxic T cells (T_C). Lyn is a tyrosine kinase which is essential for establishing immunoreceptor tyrosine-based inhibitory motif-dependent signaling and for activation of specific protein tyrosine phosphatases in myeloid cells. It is possible that the higher expression of the tyrosine kinase Lyn in CD8+CD28− T_S cells is required for establishing immunoreceptor tyrosine-based inhibitory motifs dependent signaling through KIRs.

Taken together our data demonstrate that alloantigen-specific CD8+CD28− Ts share with natural CD4+CD25+ T reg increased mRNA expression of genes associated with their suppressor function.

Molecular and Functional Events Resulting from T_S-Mediated Suppression

To define the molecular changes induced by T_S in DC, we analyzed the mRNA expression profiles of tolerogenic DC using Affymetrix gene chips. The overall picture that emerged was that tolerogenic DC differs from both immature and mature DC with respect to molecules involved in signal transduction, chemokines, cytokines, transcription factors, apoptosis-related proteins and cell growth regulators [42]. Most importantly, tolerogenic DC exhibit a high cell surface expression of the inhibitory molecules, ILT3 and ILT4, which are crucial to the tolerogenic capacity acquired by DC [41].

ILT3 and ILT4 were thought to be expressed exclusively by monocytes, macrophages and DC. They were shown to display long cytoplasmic tails containing immunoreceptor tyrosine-based inhibitory motifs and inhibit cell activation by recruiting protein tyrosine phosphatase SHP-1 [57–61].

We found that tolerogenic DC with increased ILT3 and ILT4 expression induce T_H anergy inhibiting the capacity of alloreactive CD4+ T cells to proliferate [41].

By overexpressing ILT3 and ILT4 as *myc* fusion proteins in the DC line KG1, we generated tolerogenic DC lines which had a reduced capacity to transcribe NF-κB-dependent costimulatory molecules and induced anergy in primed or unprimed CD4+ T cells. The state of anergy induced by ILT3[high] and/or ILT4[high] DC in the T_H cells can be partially abrogated by the corresponding monoclonal antibodies, thus further demonstrating the tolerogenic function of these molecules [41].

In further studies we demonstrated that CD4+CD25− T cells allostimulated with ILT3[high] ILT4[high] DC acquire a characteristic CD4$^+$+CD25+CD45RO +FOXP3+ phenotype. These alloantigen-induced T_R are MHC class II allorestricted. CD4+CD25+ T_R can be propagated in a medium containing IL-2. Similar to CD8+CD28− T_S, the anergic CD4+CD25+ T_R cells act directly on APC in a cytokine-independent manner, inducing the up-regulation of the inhibitory receptors ILT3 and ILT4. These inhibitory receptors are crucial to the tolerogenic phenotype acquired by APCs, since the suppressive effect of T_R on T_H proliferation is abrogated by mAb to ILT3 and ILT4. The finding that multiple stimulation of T cells with allogeneic APC results in the differentiation of both CD8+CD28− T_S and CD4+CD25+ T_R, yet that T_R do not develop in cultures depleted of CD8+ cells suggests that allospecific CD8+CD28− T_S initiate a 'T suppressor cell cascade' by first tolerizing the APC [47]. These tolerized APC anergize alloreactive CD4+ T_H cells, which recognize MHC class II alloantigens on their membrane. In turn, anergic CD4+CD25+ T cells act as T_R cells tolerizing other APC. Finally, APC tolerized by anergic CD4+CD25+ T_R cells may inhibit the alloreactivity of other CD4+ T_H cells, thus continuing the cascade of suppression. Central to this model is the finding that tolerized APC with up-regulated expression of the inhibitory receptors ILT3 and ILT4 spread unresponsiveness to antigen-specific T cells [47].

Such a cascade of events may provide a link between different types of T regs described before and may also explain the phenomenon of 'infectious tolerance' occurring when allograft tolerance is adoptively transferred through successive generations of naïve recipients [62]. Both stable and infectious tolerance may depend on the bidirectional interaction between T cells and tolerogenic APC which perpetuates the generation of regulatory elements.

To determine whether the up-regulation of ILT3 and ILT4 is a common characteristic of tolerogenic DC, we investigated the expression of these molecules on myeloid DC that were treated with IL-10, IFN-α and/or vitamin D3 for 24 h. Strong upregulation of ILT3 and ILT4 was induced by these agents in conjunction with the induction of tolerogenic activity. CD4+ T cells allostimulated with DC pretreated with a mixture of IL-10 and IFN-α or IL-10 and vitamin D3 showed low proliferative capacity and acquired regulatory function as they inhibited the alloreactive capacity of unprimed CD4+ T cells. This suppressive effect was partially abolished by anti-ILT3 and ILT4 mAb. Similarly, peptide-primed DC, pretreated with a mixture of IL-10 and IFN- α induced the generation of CD8+CD28− T_S, which inhibited T_H and T_C function in an antigen-specific manner. Of notice, the dominant peptide epitope recognized by T_S and T_C was found to be identical suggesting that these two distinct subsets derive from the same precursor [manuscript in preparation]. Our findings represent an important step towards the development of tolerogenic vaccines as they imply that such vaccines can be obtained by using as a vehicle ILT3[high] ILT4[high] DC.

Evaluation of the Clinical Significance and Therapeutic
Potential of Allospecific T_S

We have explored the in vivo relevance of CD8+CD28− T_S and of CD4+CD25+ T_R in recipients of heart, kidney or liver transplants [41, 44–48]. Serial studies of the phenotype displayed by T cells from heart allograft recipients demonstrated a significant increase of the CD8+CD28−CD27+ perforin-negative T cell population in rejection-free patients. This phenotype is characteristic of in vitro generated T_S as demonstrated by flow cytometry and cDNA microarray profiling [44, 56].

CD8+CD28−CD27+ T cells from these recipients inhibit up-regulation of CD80 and CD86 on CD40-ligated APC from the donor. This inhibitory effect is MHC class I allorestricted demonstrating the antigen specificity of the in vivo generated T_S [45, 46]. We demonstrated that CD8+CD28− T_S from patients in quiescence induce the up-regulation of ILT3 and ILT4 on donor APC in an MHC class I-allorestricted manner [41]. CD8+CD28− T_S were found in patients' circulation within the first 6 months post-transplantation and persisted thereafter in recipients with no evidence of chronic rejection 3 years following transplantation.

Serial determination of the capacity of patient's CD4+CD25+ T cells to induce the up-regulation of ILT3 and ILT4 on donor APC in an MHC class II-allorestricted manner yielded similar results [47]. However, CD4+CD25+ T_R became detectable at later times following transplantation, that is, 3 months or more after CD8+CD28− T_S could be seen. The delayed differentiation of T_R may represent the in vivo counterpart of the T suppressor cell cascade, which is initiated by CD8+ T_S and continued by CD4+ T_R cells in vitro [47]. Further studies are required, however, for understanding the dynamics of these events.

The persistence of allospecific-T_S and -T_R in rejection-free patients late after transplantation indicates that these T cells which inhibit the direct allorecognition pathway are continuously stimulated by donor APC. However, donor DC migrate out of the graft during the early post-transplantation period, raising some questions about the identity of the APC which stimulate T_S and T_R with direct allorecognition capacity. We postulated that endothelial cells (EC), which are known to act as semi-professional APC, stimulate the direct allorecognition pathway throughout the lifetime of the graft.

We have explored this hypothesis by first analyzing the effect of CD8+CD28− T cells from allospecific TCL on human umbilical cord vein EC (HUVEC) and aortic EC (HAEC) exposed to inflammatory (IFN-γ and TNF-α) cytokines [48]. T_S induced the down-regulation of CD40, CD54, CD58, CD62E, CD106, HLA class I and HLA class II on activated EC, while concomitantly up-regulating the expression of the inhibitory receptors ILT3 and ILT4 in an HLA class I-allorestricted manner. T_H reactivity to HLA-DR+ EC was inhibited in the presence of T_S. This effect was due to up-regulation of

ILT3 and ILT4 on EC since the T$_S$ effect was abrogated by adding to the cultures of mAb to ILT3 and/or ILT4 [48]. Similar to DC, EC which had been tolerized by exposure to IFN-α and IL-10 elicited the differentiation of CD8+CD28− T$_S$ from unprimed populations of CD8+ T cells [48]. Hence, modulation of ILT3 and ILT4 expression on professional and nonprofessional APC renders these cells tolerogenic.

Based on this finding we studied the possibility that heart allograft recipients in quiescence display CD8+CD28− FOX P3+ T$_S$ which recognize specifically donor HLA class I antigens and inhibit the direct allorecognition pathway by inducing ILT3 and ILT4 expression on graft EC. For this, a panel of HUVEC lines representative of a wide array of HLA A and B alleles was transfected with pGL3 constructs containing 766 bp of the ILT4 promoter and 1034 bp of the ILT3 promoter upstream of the luciferase reporter gene. The transfected EC were used as targets for measuring the capacity of T$_S$ to induce ILT3 and ILT4 transcription. In patients tested within 10–12 months after heart transplantation, quiescence was associated with the presence in the circulation of CD8+CD28− FOXP3+ T$_S$ cells which triggered ILT3 and ILT4 transcription in donor-matched EC. A similar correlation was found in rejection-free patients tested 3 years following transplantation [48].

Conclusion

Study of the capacity of CD8+CD28− FOXP3+ T cells from recipients' circulation to induce the up-regulation of inhibitory receptors in EC, in an alloantigen-specific manner, may permit the identification of patients who will benefit from partial or complete withdrawal of immunosuppression. This is an important aim in view of the morbidity and mortality associated with the long-term use of immunosuppressive drugs. Furthermore, the development of pharmaceutical agents that can act on DC and/or EC by up-regulating inhibitory receptors, such as ILT3 and ILT4, may permit modulation of the immune response in allograft recipients and patients with autoimmune diseases. The recent finding that in vitro-generated tolerogenic APC induce CD8+ T reg cells which can suppress ongoing experimental autoimmune encephalomyelitis [63] supports the rationale for developing such new therapeutic strategies.

References

1   Hoffmann MW, Heath WR, Ruschmeyer D, Miller JF: Deletion of high-avidity T cells by thymic epithelium. Proc Natl Acad Sci USA 1995;92:9851–9855.
2   Janssen O, Sanzenbacher R, Kabelitz D: Regulation of activation-induced cell death of mature T-lymphocyte populations. Cell Tissue Res 2000;301:85–99.

3   Nguyen T, Russell J: The regulation of FasL expression during activation-induced cell death (AICD). Immunology 2001;103:426–434.

4   Conroy LA, Alexander DR: The role of intracellular signalling pathways regulating thymocyte and leukemic T cell apoptosis. Leukemia 1996;10:1422–1435.

5   Oberg HH, Lengl-Janssen B, Kabelitz D, Janssen O: Activation-induced T cell death: Resistance or susceptibility correlate with cell surface fas ligand expression and T helper phenotype. Cell Immunol 1997;181:93–100.

6   Oberg HH, Sanzenbacher R, Lengl-Janssen B, Dobmeyer T, Flindt S, Janssen O, Kabelitz D: Ligation of cell surface CD4 inhibits activation-induced death of human T lymphocytes at the level of Fas ligand expression. J Immunol 1997;159:5742–5749.

7   Lee KM, Chuang E, Griffin M, Khattri R, Hong DK, Zhang W, Straus D, Samelson LE, Thompson CB, Bluestone JA: Molecular basis of T cell inactivation by CTLA-4. Science 1998;282:2263–2266.

8   Wells AD, Walsh MC, Bluestone JA, Turka LA: Signaling through CD28 and CTLA-4 controls two distinct forms of T cell anergy. J Clin Invest 2001;108:895–903.

9   Chambers CA, Allison JP: Costimulatory regulation of T cell function. Curr Opin Cell Biol 1999;11:203–210.

10   Suciu-Foca N, Manavalan JS, Cortesini R: Generation and function of antigen-specific suppressor and regulatory T cells. Transpl Immunol 2003;11:235–244.

11   Sutmuller RP, Offringa R, Melief CJ: Revival of the regulatory T cell: New targets for drug development. Drug Discov Today 2004;9:310–316.

12   Nagler-Anderson C, Bhan AK, Podolsky DK, Terhorst C: Control freaks: Immune regulatory cells. Nat Immunol 2004;5:119–122.

13   Sakaguchi S, Sakaguchi N, Asano M, Itoh M, Toda M: Immunologic self-tolerance maintained by activated T cells expressing IL-2 receptor alpha-chains (CD25). Breakdown of a single mechanism of self-tolerance causes various autoimmune diseases. J Immunol 1995;155:1151–1164.

14   Shevach EM, McHugh RS, Piccirillo CA, Thornton AM: Control of T-cell activation by CD4+ CD25+ suppressor T cells. Immunol Rev 2001;182:58–67.

15   Sakaguchi S, Sakaguchi N, Shimizu J, Yamazaki S, Sakihama T, Itoh M, Kuniyasu Y, Nomura T, Toda M, Takahashi T: Immunologic tolerance maintained by CD25+ CD4+ regulatory T cells: Their common role in controlling autoimmunity, tumor immunity, and transplantation tolerance. Immunol Rev 2001;182:18–32.

16   Itoh M, Takahashi T, Sakaguchi N, Kuniyasu Y, Shimizu J, Otsuka F, Sakaguchi S: Thymus and autoimmunity: Production of CD25+CD4+ naturally anergic and suppressive T cells as a key function of the thymus in maintaining immunologic self-tolerance. J Immunol 1999;162:5317–5326.

17   Shevach EM: Regulatory T cells in autoimmmunity*. Annu Rev Immunol 2000;18:423–449.

18   Hori S, Nomura T, Sakaguchi S: Control of regulatory T cell development by the transcription factor Foxp3. Science 2003;299:1057–1061.

19   Takahashi T, Kuniyasu Y, Toda M, Sakaguchi N, Itoh M, Iwata M, Shimizu J, Sakaguchi S: Immunologic self-tolerance maintained by CD25+CD4+ naturally anergic and suppressive T cells: Induction of autoimmune disease by breaking their anergic/suppressive state. Int Immunol 1998;10:1969–1980.

20   Thornton AM, Donovan EE, Piccirillo CA, Shevach EM: Cutting edge: IL-2 is critically required for the in vitro activation of CD4+CD25+ T cell suppressor function. J Immunol 2004;172: 6519–6523.

21   Jordan MS, Boesteanu A, Reed AJ, Petrone AL, Holenbeck AE, Lerman MA, Naji A, Caton AJ: Thymic selection of CD4+CD25+ regulatory T cells induced by an agonist self-peptide. Nat Immunol 2001;2:301–306.

22   Walker MR, Kasprowicz DJ, Gersuk VH, Benard A, Van Landeghen M, Buckner JH, Ziegler SF: Induction of FoxP3 and acquisition of T regulatory activity by stimulated human CD4+CD25- T cells. J Clin Invest 2003;112:1437–1443.

23   Read S, Malmstrom V, Powrie F: Cytotoxic T lymphocyte-associated antigen 4 plays an essential role in the function of CD25(+)CD4(+) regulatory cells that control intestinal inflammation. J Exp Med 2000;192:295–302.

24   Kingsley CI, Karim M, Bushell AR, Wood KJ: CD25+CD4+ regulatory T cells prevent graft rejection: CTLA-4- and IL-10-dependent immunoregulation of alloresponses. J Immunol 2002; 168:1080–1086.

25 Shimizu J, Yamazaki S, Takahashi T, Ishida Y, Sakaguchi S: Stimulation of CD25(+)CD4(+) regulatory T cells through GITR breaks immunological self-tolerance. Nat Immunol 2002;3:135–142.

26 McHugh RS, Whitters MJ, Piccirillo CA, Young DA, Shevach EM, Collins M, Byrne MC: CD4(+)CD25(+) Immunoregulatory T cells: Gene expression analysis reveals a functional role for the glucocorticoid-induced TNF receptor. Immunity 2002;16:311–323.

27 Groux H, O'Garra A, Bigler M, Rouleau M, Antonenko S, de Vries JE, Roncarolo MG: A CD4+ T-cell subset inhibits antigen-specific T-cell responses and prevents colitis. Nature 1997;389:737–742.

28 Bacchetta R, Bigler M, Touraine JL, Parkman R, Tovo PA, Abrams J, de Waal Malefyt R, de Vries JE, Roncarolo MG: High levels of interleukin 10 production in vivo are associated with tolerance in SCID patients transplanted with HLA mismatched hematopoietic stem cells. J Exp Med 1994;179:493–502.

29 Asseman C, Mauze S, Leach MW, Coffman RL, Powrie F: An essential role for interleukin 10 in the function of regulatory T cells that inhibit intestinal inflammation. J Exp Med 1999;190:995–1004.

30 Kitani A, Chua K, Nakamura K, Strober W: Activated self-MHC-reactive T cells have the cytokine phenotype of Th3/T regulatory cell 1 T cells. J Immunol 2000;165:691–702.

31 Powrie F, Carlino J, Leach MW, Mauze S, Coffman RL: A critical role for transforming growth factor-beta but not interleukin 4 in the suppression of T helper type 1-mediated colitis by CD45RB(low) CD4+ T cells. J Exp Med 1996;183:2669–2674.

32 Levings MK, Sangregorio R, Galbiati F, Squadrone S, de Waal Malefyt R, Roncarolo MG: IFN-alpha and IL-10 induce the differentiation of human type 1 T regulatory cells. J Immunol 2001;166:5530–5539.

33 Dieckmann D, Bruett CH, Ploettner H, Lutz MB, Schuler G: Human CD4(+)CD25(+) regulatory, contact-dependent T cells induce interleukin 10-producing, contact-independent type 1-like regulatory T cells [corrected]. J Exp Med 2002;196:247–253.

34 Jonuleit H, Schmitt E, Kakirman H, Stassen M, Knop J, Enk AH: Infectious tolerance: Human CD25(+) regulatory T cells convey suppressor activity to conventional CD4(+) T helper cells. J Exp Med 2002;196:255–260.

35 Liu Z, Tugulea S, Cortesini R, Suciu-Foca N: Specific suppression of T helper alloreactivity by allo-MHC class I-restricted CD8+CD28− T cells. Int Immunol 1998;10:775–783.

36 Ciubotariu R, Colovai AI, Pennesi G, Liu Z, Smith D, Berlocco P, Cortesini R, Suciu-Foca N: Specific suppression of human CD4+ Th cell responses to pig MHC antigens by CD8+CD28− regulatory T cells. J Immunol 1998;161:5193–5202.

37 Jiang S, Tugulea S, Pennesi G, Liu Z, Mulder A, Lederman S, Harris P, Cortesini R, Suciu-Foca N: Induction of MHC-class I restricted human suppressor T cells by peptide priming in vitro. Hum Immunol 1998;59:690–699.

38 Colovai AI, Liu Z, Ciubotariu R, Lederman S, Cortesini R, Suciu-Foca N: Induction of xenoreactive CD4+ T-cell anergy by suppressor CD8+CD28− T cells. Transplantation 2000;69:1304–1310.

39 Liu Z, Tugulea S, Cortesini R, Lederman S, Suciu-Foca N: Inhibition of CD40 signaling pathway in antigen presenting cells by T suppressor cells. Hum Immunol 1999;60:568–574.

40 Li J, Liu Z, Jiang S, Cortesini R, Lederman S, Suciu-Foca N: T suppressor lymphocytes inhibit NF-kappa B-mediated transcription of CD86 gene in APC. J Immunol 1999;163:6386–6392.

41 Chang CC, Ciubotariu R, Manavalan JS, Yuan J, Colovai AI, Piazza F, Lederman S, Colonna M, Cortesini R, Dalla-Favera R, Suciu-Foca N: Tolerization of dendritic cells by T(S) cells: The crucial role of inhibitory receptors ILT3 and ILT4. Nat Immunol 2002;3:237–243.

42 Suciu-Foca Cortesini N, Piazza F, Ho E, Ciubotariu R, LeMaoult J, Dalla-Favera R, Cortesini R: Distinct mRNA microarray profiles of tolerogenic dendritic cells. Hum Immunol 2001;62: 1065–1072.

43 Cortesini R, LeMaoult J, Ciubotariu R, Cortesini NS: CD8+CD28− T suppressor cells and the induction of antigen-specific, antigen-presenting cell-mediated suppression of Th reactivity. Immunol Rev 2001;182:201–206.

44 Colovai AI, Mirza M, Vlad G, Wang S, Ho E, Cortesini R, Suciu-Foca N: Regulatory CD8+CD28− T cells in heart transplant recipients. Hum Immunol 2003;64:31–37.

45 Ciubotariu R, Vasilescu R, Ho E, Cinti P, Cancedda C, Poli L, Late M, Liu Z, Berloco P, Cortesini R, Suciu-Foca Cortesini N: Detection of T suppressor cells in patients with organ allografts. Hum Immunol 2001;62:15–20.

46 Cortesini R, Renna-Molajoni E, Cinti P, Pretagostini R, Ho E, Rossi P, Suciu-Foca Cortesini N: Tailoring of immunosuppression in renal and liver allograft recipients displaying donor specific T-suppressor cells. Hum Immunol 2002;63:1010–1018.

47 Manavalan JS, Rossi PC, Vlad G, Piazza F, Yarilina A, Cortesini R, Mancini D, Suciu-Foca N: High expression of ILT3 and ILT4 is a general feature of tolerogenic dendritic cells. Transpl Immunol 2003;11:245–258.

48 Manavalan JS, Kim-Schulze S, Scotto L, Naiyer AJ, Vlad G, Colombo P, Marboe C, Mancini D, Cortesini R, Suciu-Foca N: Alloantigen specific CD8+ CD28− FOXP3+ T suppressors cells induce tolerogenic endothelial cells, inhibiting alloreactivity. Int Immunol DOI:10.1159/000063858.

49 Stephens LA, Mason D: CD25 is a marker for CD4+ thymocytes that prevent autoimmune diabetes in rats, but peripheral T cells with this function are found in both CD25+ and CD25− subpopulations. J Immunol 2000;165:3105–3110.

50 Shimizu J, Yamazaki S, Takahashi T, Ishida Y, Sakaguchi S: Stimulation of CD25(+)CD4(+) regulatory T cells through GITR breaks immunological self-tolerance. Nat Immunol 2002;3:135–142.

51 Fu S, Yopp AC, Mao X, Chen D, Zhang N, Mao M, Ding Y, Bromberg JS: CD4+ CD25+ CD62+ T-regulatory cell subset has optimal suppressive and proliferative potential. Am J Transplant 2004;4:65–78.

52 Takeda I, Ine S, Killeen N, Ndhlovu LC, Murata K, Satomi S, Sugamura K, Ishii N: Distinct roles for the OX40-OX40 ligand interaction in regulatory and nonregulatory T cells. J Immunol 2004;172:3580–3589.

53 Choi BK, Bae JS, Choi EM, Kang WJ, Sakaguchi S, Vinay DS, Kwon BS: 4–1BB-dependent inhibition of immunosuppression by activated CD4+CD25+ T cells. J Leukoc Biol 2004;75:785–791.

54 Annunziato F, Cosmi L, Liotta F, Lazzeri E, Manetti R, Vanini V, Romagnani P: Phenotype, localization, and mechanism of suppression of CD4(+)CD25(+) human thymocytes. J Exp Med 2002; 196:379–387.

55 Lehmann J, Huehn J, de la Rosa M, Maszyna F, Kretschmer U, Krenn V, Brunner M, Scheffold A, Hamann A: Expression of the integrin alpha Ebeta 7 identifies unique subsets of CD25+ as well as CD25− regulatory T cells. Proc Natl Acad Sci USA 2002;99:13031–13036.

56 Suciu-Foca N, Manavalan JS, Scotto L, Kim-Schulze S, Galluzzo S, Naiyer AJ, Fan J, Vlad G, Cortesini R: Molecular characterization of allospecific T suppressor and tolerogenic dendritic cells: Review. Int J Immunopharmacol; in press.

57 Colonna M, Nakajima H, Navarro F, Lopez-Botet M: A novel family of Ig-like receptors for HLA class I molecules that modulate function of lymphoid and myeloid cells. J Leukoc Biol 1999; 66:375–381.

58 Colonna M, Nakajima H, Cella M: A family of inhibitory and activating Ig-like receptors that modulate function of lymphoid and myeloid cells. Semin Immunol 2000;12:121–127.

59 Colonna M, Samaridis J, Cella M, Angman L, Allen RL, O'Callaghan CA, Dunbar R, Ogg GS, Cerundolo V, Rolink A: Human myelomonocytic cells express an inhibitory receptor for classical and nonclassical MHC class I molecules. J Immunol 1998;160:3096–3100.

60 Ravetch JV, Lanier LL: Immune inhibitory receptors. Science 2000;290:84–89.

61 Cella M, Dohring C, Samaridis J, Dessing M, Brockhaus M, Lanzavecchia A, Colonna M: A novel inhibitory receptor (ILT3) expressed on monocytes, macrophages, and dendritic cells involved in antigen processing. J Exp Med 1997;185:1743–1751.

62 Waldmann H, Cobbold S: Regulating the immune response to transplants. A role for CD4+ regulatory cells? Immunity 2001;14:399–406.

63 Najafian N, Chitnis T, Salama AD, Zhu B, Benou C, Yuan X, Clarkson MR, Sayegh MH, Khoury SJ: Regulatory functions of CD8+CD28− T cells in an autoimmune disease model. J Clin Invest 2003;112:1037–1048.

Prof. Nicole Suciu-Foca
Columbia University
630 West 168 Street – P&S 14–401
New York, NY 10032 (USA)
Tel. +1 212 305–6941, Fax +1 212 305–3429, E-Mail ns20@columbia.edu

Ronco C, Chiaramonte S, Remuzzi G (eds): Kidney Transplantation: Strategies to Prevent
Organ Rejection. Contrib Nephrol. Basel, Karger, 2005, vol 146, pp 143–150

....

The Goal of Intragraft Gene Therapy

Susanna Tomasoni, Ariela Benigni

Mario Negri Institute for Pharmacological Research, Bergamo, Italy

---

Abstract

Despite the impressive results of one-year survival rates, organ transplantation still
faces major problems. Current anti-rejection drugs reduce systemic immunity nonselectively
and increase the risk of infection and cancer on the long term. Theoretically, selective
inhibition of alloimmune response can be achieved at the organ level by intragraft transfer of
genes with immunomodulatory properties. In the last decade, gene therapy emerged as a
new strategy in renal, heart and liver transplantation, showing promising results in experimental animals, almost in controlling acute rejection. The success of gene therapy in the
transplant medicine is strongly dependent on the efficiency of the delivery system that
allows local transfer and expression of the therapeutic gene in the target organ or tissue. The
main findings concerning the suitability of gene therapy in preventing graft rejection will be
discussed here.

Introduction

Since the success in generating transgenic animals, techniques leading to
the manipulation of the mammalian genome have provided pivotal information
on the role of gene products in vivo. From the genetic engineering methodol-
ogy derived the interest for gene therapy as a tool to treat human diseases
through the introduction of foreign genetic material. Moreover, gene therapy
may find widespread application for the treatment of acquired diseases characterized by low or high expression of a given protein. However, several practical
hurdles have reduced the enthusiasm for gene therapy as an immediate
perspective for genetic and acquired disorders requiring targeting delivery to a
specific cell type in vivo. Instead, efforts to identify clinical settings suitable
for ex-vivo genetic delivery to a given organ or tissue has credited it as a new
strategy, particularly in solid organ transplantation and cell therapy. The

success of gene therapy largely depends on the efficiency of the delivery system to transfer and express the therapeutic gene. An ideal gene therapy vector should be nontoxic, nonimmunogenic, easy to produce in large quantities, and efficient in protecting and delivering DNA into cells, preferably to a specific target cell. However, a variety of barriers exists that can limit the efficiency of the gene transfer approach. Once the extracellular barriers to the virus' entry into the cells have been overcome, the vector faces cellular barriers, such as nucleases and endosomal entrapment, hurdles that limit the access of the nucleic acids to the nucleus. In addition, host innate and adaptive immune response against the vector proteins and the transgene product still impair the gene transfer efficiency and, moreover, impede a second viral administration. This review will be focused on the progress reached in the last few years in preventing transplant rejection by using an intragraft gene therapy approach.

Intragraft Gene Therapy

The opportunity to perform ex vivo manipulation of the graft during organ retrieval makes transplantation an ideal condition to achieve local immunosuppression. In the last decade, it has been shown that most of the transplantable solid organs such as kidney, liver and heart are receptive to gene transfer by the currently available vectors. However, the efficiency of gene transfer can strongly vary depending on the vector used, the biological characteristics of the targeted cells as well as on the activation of the host immune system in response to the vector and to the transgene product. Indeed, most of the vectors induce an innate immune response that can impair the efficiency of this approach. This is particularly true when adenovirus, of any generation, is used. Different bioactive molecules have been delivered to the donor organ with the aim of prolonging survival and, hopefully, inducing tolerance of the graft. Specifically, three different strategies have been pursued: (1) blocking the anti-graft immune responses by inhibiting the costimulatory signals or inducing apoptosis of immune cells, (2) inducing the local production of immunomodulatory cytokines such as interleukin (IL)-10 or transforming growth factor-β (TGF-β), or (3) protecting the donor organ against the host immune responses by the overexpression of anti-apoptotic genes (table 1).

Inhibition of Costimulatory Pathways and Induction of Activated T Cell Apoptosis
Graft rejection is a consequence of full T cell activation, a process involving engagement of T cell receptor and alloantigens presented by MHC molecules on the surface of the antigen presenting cells, and needing costimulatory

Table 1. Transgene used to engineer the graft

Transgene	Action	Allograft	Reference
CTLA4Ig	Inhibition of CD28/B7 costimulatory pathway	Kidney Liver Heart	[2] [3] [4, 5]
CD40Ig	Inhibition of CD40/CD154 costimulatory pathway	Liver Heart	[6] [7]
CTLA4Ig + CD40Ig	Inhibition of both costimulatory pathways	Heart (systemic injection)	[9]
FasL	T cell apoptosis	Kidney	[10, 11]
vIL-10	Induction of immunomodulation	Heart Kidney Liver	[12–15] [16] [17]
IL-13 ± regulatory T cells		Heart	[18]
TGF-β		Heart Kidney	[19, 20] [16]
vIL-10 + TNFRp55-Ig + IL-12p40		Kidney	[21]
SOD	Anti-oxidative stress	Liver	[22]
HO-1		Liver Heart	[23, 24] [25]

signals [1]. One of the most characterized costimulatory pathways is the binding between CD28 (on the T cell) and B7 (on the antigen presenting cell) that, following T cell receptor ligation, induces T cell activation and expansion. Blocking this costimulatory signal through the chimeric fusion protein cytotoxic T lymphocyte antigen-4 Ig (CTLA4Ig) that binds B7 with high affinity, immune responses are down-regulated as shown by several experimental evidences. Direct injection of a recombinant adenovirus encoding CTLA4Ig into the renal artery of the donor kidney before transplantation significantly prolonged graft survival of cold preserved rat renal allografts without the need of systemic immunosuppression [2]. Despite the mild infiltration of mononuclear cells which was observed in the transfected organs, renal function was well preserved in the long lasting animals that showed donor-specific unresponsiveness. Survival was indefinite in rat recipients of cold preserved liver transduced with AdCTLA4Ig that developed donor-specific unresponsiveness [3]. A single ex vivo intra-arterial infusion of recombinant adenovirus encoding CTLA4Ig induced efficient transduction of the endomyocardium promoting permanent acceptance of cardiac allografts in nonimmunosuppressed rats [4]. Similar results were obtained when CTLA4Ig was injected directly into the myocardium [5].

A second costimulatory pathway important for initiation and maintenance of T cell responses is CD40, expressed on antigen presenting cells, and its ligand CD154 expressed on T lymphocytes. Similarly to CTLA4Ig gene therapy, CD40Ig fusion protein is able to block the CD40/CD154 interaction. Targeted gene therapy with adenovirus encoding CD40Ig has been attempted in liver and heart transplantation. Rat liver allografts transduced before transplantation survived more than 100 days displaying normal histology. Donor specific unresponsiveness was demonstrated by survival of a second skin allograft [6]. Blockade of CD40/CD154 interaction by adenoviral-mediated gene transfer of CD40Ig resulted in long-term heart allograft survival and induced donor-specific hyporesponsiveness. However, despite strong inhibition of alloantibody production and allogeneic proliferation, signs of chronic rejection were detected in the long-surviving grafts [7].

A combination therapy of CTLA4Ig and CD40Ig gene transfer has been attempted with the aim of inducing systemic immunosuppression. Recently, systemic administration of both adenoviruses prolonged survival of rat hind limb allografts to a longer extent than the single therapy [8]. Very interesting data have been shown in the context of heart transplantation. Intravenous injection of both adenoviruses induced long-term acceptance (more than 270 days) of rat cardiac allografts. However, this strategy was not enough to allow tolerance to a second allograft and to avoid chronic rejection in long-term surviving heart grafts [9]. At the best of our knowledge, a local intragraft combined approach has not yet been attempted.

Induction of apoptosis of alloreactive T cells may represent an alternative way to prolong graft survival. Activated T cells are usually eliminated by apoptosis, triggered by the interaction between the Fas antigen and its counter receptor Fas-ligand (FasL). Theoretically, expression of FasL within the graft should protect it from infiltrating T cells that, by expressing Fas antigen, should undergo apoptosis. Renal allografts adenovirally transduced to express FasL showed prolonged graft survival, an effect correlating with a down-regulation of Bag-1 and enhanced T helper-type 2 cytokines (IL-4 and IL-10) mRNA expression [10, 11].

Induction of Immunomodulatory Cytokines

Graft tolerance is usually associated with a decreased production of T helper-type 1 cytokines and/or increased production of T helper-type2-derived cytokines IL-4, IL-10, IL-13 or TGF-β. Intragraft overexpression of immunomodulatory cytokines has shown quite positive results in promoting graft tolerance.

Prolonged survival of cardiac allografts has been obtained after viral IL-10 (vIL-10) gene transfer that shares some biological activities of cellular IL-10 but lacks the immunostimulatory functions, making it a potentially potent immunosuppressant. Different viral and nonviral approaches have been evaluated in heart transplantation. Prolonged but not indefinite survival was obtained after retroviral [12], adenoviral [13] or lipid [14]-mediated gene transfer of vIL-10 in cardiac allografts before transplantation. However, high levels of expression of vIL-10 is not always beneficial. Indeed, heart grafts from transgenic mice for vIL-10 failed to exhibit prolonged survival when transplanted in MHC full-mismatched animals [15]. Very recently, we evaluated whether adenoviral gene transfer of vIL-10 into the rat donor kidney before transplantation was able to prolong survival of the graft. However, only two out of nine animals treated with vIL-10 survived longer than control animals [16]. Better results have been shown in the context of liver transplantation using an adenovirus encoding for human IL-10. Intraportal injection in the donor organ of the adenovirus, 24–48 h before transplantation, induced a significant prolongation of graft survival (more than 87 days) [17]. Of note, in this latter study the experimental setting was different from the previous ones where transfection was performed at the time of transplantation; moreover, the efficiency of adenoviral gene transfer is normally higher in the liver than in other organs and yet, liver grafts are considered to be more tolerant than heart or kidney.

Gene transfer of IL-13 in cardiac allograft induced a modest prolongation of the graft survival, effect that was enhanced combining the gene transfer approach with adoptive transfer of regulatory T cells. Local IL-13 diminished intragraft apoptosis, and up-regulated anti-apoptotic A20 and anti-oxidant heme oxygenase 1 (HO-1) [18].

TGF-β gene transfer has also been attempted. Heart allografts showed prolonged survival after intragraft TGF-β1 gene transfer [19, 20]. However, signs of chronic rejection were detected in the long-term allografts [20]. Our group found that TGF-β3 gene transfer was not consistently effective in prolonging kidney allograft survival but, when combined with the adenovirus encoding for CTLA4Ig, survival was significantly prolonged in all animals, one of which survived indefinitely (more than 263 days) [16].

Although the exact mechanisms responsible for chronic allograft rejection are still not well understood, attempts have been made to overcome it in the context of renal transplantation. Recently, it has been shown that chronic renal injury could be relieved combining different adenoviral constructs expressing vIL-10, the chimeric molecule TNF receptor-Ig and IL-12p40, the beneficial effect correlating with less macrophage infiltration. By contrast, intragraft overexpression of IFN-γ-accelerated chronic graft rejection [21].

Improving Graft Function by Cytoprotective Genes

The anoxia/ischemia-reperfusion injury that occurs in the donor organ at the time of transplantation is a critical factor in conditioning the function of the graft even in the long term. Thus, transfer into the organ of genes encoding for molecules with protective actions may preserve the donor organ from this insult.

Oxygen-derived free radicals produced during the ischemic damage are responsible for cell death in the graft. Endogenous scavengers, such as super-oxide dismutase (SOD), are able to degrade toxic radicals. On this basis, an intragraft overexpression of gene encoding SOD could be beneficial for the graft. This strategy has been attempted in liver transplantation. Adenoviral transduction of the liver with the Cu/Zn-SOD gene allowed 100% survival of transplanted animals while only 20–25% of animals treated with an irrelevant adenovirus survived [22].

HO-1 is the inducible HO isoform with cytoprotective effects against the oxidative stress. Overexpression of HO-1 in the donor rat liver before trans-plantation significantly increased survival of treated grafts and improved liver function, decreased macrophage infiltration and increased intragraft expression of the anti-apoptotic genes, Bcl-2 and Bag-1 [23, 24]. Similar data have been obtained in cardiac allograft. The intragraft injection of adenovirus encoding for HO-1 as well as intramuscular and intravenous administration, prolonged allograft survival, an effect associated with inhibition of allogeneic cellular immune responses [25].

Conclusions

From the experimental data obtained in solid organ transplantation, it emerges that intragraft gene transfer could represent a promising tool to avoid or at least reduce the need for systemic immunosuppression and its deleterious consequences. However, many hurdles remain to be overcome. Extensive work has to be done in rendering the vectors more safe and less immunogenic. Safety is a problem particularly true for integrating vectors, such as retrovirus – so far the most widely used in clinical trials – that could lead to activation of proto-oncogenes or inactivation of tumor suppressor genes. On the other hand, vectors like adenovirus, the most efficient in delivering the genetic material into the nucleus, are the most immunogenic of all vector types, inducing strong innate and adaptive immune responses. A fundamental requisite to succeed in gene therapy is the possibility to modulate and regulate the transgene expression for the appropriate length of time. Moreover, a detailed understanding of the trans-plant immunobiology will be necessary to target the specific pathways involved in acute and chronic allograft rejection. Data derived from the preclinical studies represent a good premise to the use of intragraft gene therapy, possibly combined with low dose immunosuppressant or injection of regulatory T cells, as a therapeutic strategy for preventing allograft rejection in a short while.

References

1   Perico N, Remuzzi G: Prevention of transplant rejection. Current treatment guidelines and future developments. Drugs 1997;54:533–570.
2   Tomasoni S, Azzollini N, Casiraghi F, Capogrossi MC, Remuzzi G, Benigni A: CTLA4Ig gene transfer prolongs survival and induces donor-specific tolerance in a rat renal allograft. J Am Soc Nephrol 2000;11:747–752.
3   Olthoff KM, Judge TA, Gelman AE, Da Shen X, Hancock WW, Turka LA, Shaked A: Adenovirus-mediated gene transfer into cold-preserved liver allografts: Survival pattern and unresponsiveness following transduction with CTLA4Ig. Nat Med 1998;4:194–200.
4   Yang Z, Rostami S, Koeberlein B, Barker CF, Naji A: Cardiac allograft tolerance induced by intra-arterial infusion of recombinant adenoviral CTLA4Ig. Transplantation 1999;67:1517–1523.
5   Guillot C, Mathieu P, Coathalem H, Gerdes CA, Menoret S, Braudeau C, Tesson L, Renaudin K, Castro MG, Lowenstein PR, Anegon I: Tolerance to cardiac allografts via local and systemic mechanisms after adenovirus-mediated CTLA4Ig expression. J Immunol 2000;164:5258–5268.
6   Chang GJ, Liu T, Feng S, Bedolli M, O'rourke RW, Schmidt G, Roberts JP, Stock PG: Targeted gene therapy with CD40Ig to induce long-term acceptance of liver allografts. Surgery 2002; 132:149–156.
7   Guillot C, Guillonneau C, Mathieu P, Gerdes CA, Menoret S, Braudeau C, Tesson L, Renaudin K, Castro MG, Lowenstein PR, Anegon I: Prolonged blockade of CD40-CD40 ligand interactions by gene transfer of CD40Ig results in long-term heart allograft survival and donor-specific hypore-sponsiveness, but does not prevent chronic rejection. J Immunol 2002;168:1600–1609.
8   Kanaya K, Tsuchida Y, Inobe M, Murakami M, Hirose T, Kon S, Kawaguchi S, Wada T, Yamashita T, Ishii S, Uede T: Combined gene therapy with adenovirus vectors containing CTLA4Ig and CD40Ig prolongs survival of composite tissue allografts in rat model. Transplantation 2003;75: 275–281.
9   Yamashita K, Masunaga T, Yanagida N, Takehara M, Hashimoto T, Kobayashi T, Echizenya H, Hua N, Fujita M, Murakami M, Furukawa H, Uede T, Todo S: Long-term acceptance of rat cardiac allografts on the basis of adenovirus mediated CD40Ig plus CTLA4Ig gene therapies. Transplantation 2003;76:1089–1096.
10 Swenson KM, Ke B, Wang T, Markowitz JS, Maggard MA, Spear GS, Imagawa DK, Goss JA, Busuttil RW, Seu P: Fas ligand gene transfer to renal allografts in rats. Transplantation 1998; 65:155–160.
11 Ke B, Coito AJ, Kato H, Zhai Y, Wang T, Sawatzki B, Seu P, Busuttil RW, Kupiec-Weglinski JW: Fas ligand gene transfer prolongs rat renal allograft survival and down-regulates anti-apoptotic bag-1 in parallel with enhanced Th2-type cytokine expression. Transplantation 2000;69: 1690–1694.
12 Qin L, Chavin KD, Ding Y, Tahara H, Favaro JP, Woodward JE, Suzuki T, Robbins PD, Lotze MT, Bromberg JS: Retrovirus-mediated transfer of viral IL-10 gene prolongs murine cardiac allograft survival. J Immunol 1996;156:2316–2323.
13 Zuo Z, Wang C, Carpenter D, Okada Y, Nicolaidou E, Yoyoda M, Trento A, Jordan SC: Prolongation of allograft survival with viral IL-10 transfection in a highly histoincompatible model of rat heart allograft rejection. Transplantation 2001;71:686–691.
14 DeBruyne LA, Li K, Chan SY, Qin L, Bishop DK, Bromberg JS: Lipid-mediated gene transfer of viral IL-10 prolongs vascularized cardiac allograft survival by inhibiting donor-specific cellular and humoral immune responses. Gene Ther 1998;5:1079–1087.

15 Adachi O, Yamato E, Kawamoto S, Yamamoto M, Tahara H, Tabayashi K, Miyazaki J-I: High-level expression of viral interleukin-10 in cardiac allografts fails to prolong graft survival. Transplantation 2002;74:1603–1608.

16 Tomasoni S, Longaretti L, Azzollini N, Gagliardini E, Mister M, Buehler T, Remuzzi G, Benigni A: Favorable effect of cotransfection with TGF-beta and CTLA4Ig of the donor kidney on allograft survival. Am J Nephrol 2004;24:275–283.

17 Shinozaki K, Yahata H, Tanji H, Sakaguchi T, Ito H, Dohi K: Allograft transduction of IL-10 prolongs survival following orthotopic liver transplantation. Gene Ther 1999;6:816–822.

18 Ke B, Shen XD, Zhai Y, Gao F, Busuttil RW, Volk H-D, Kupiec-Weglinski JW: Heme oxygenase 1 mediates the immunomodulatory and antiapoptotic effects of interleukin 13 gene therapy in vivo and in vitro. Hum Gene Ther 2002;13:1845–1857.

19 Qin L, Chavin KD, Ding Y, Favaro JP, Woodward JE, Lin J, Tahara H, Robbins PD, Shaked A, Ho DY, Sapolsky RM, Lotze MT, Bromberg JS: Multiple vectors effectively achieve gene transfer in a murine cardiac transplantation model. Transplantation 1995;59:809–816.

20 Chan SY, Goodman RE, Szmuszkovicz JR, Roessler B, Eichwald EJ, Bishop K: DNA-liposome versus adenoviral mediated gene transfer of transforming growth factor beta-1 in vascularized cardiac allografts: Differential sensitivity of CD4+ and CD8+ T cells to transforming growth factor beta-1. Transplantation 2000;70:1292–1301.

21 Yang J, Reutzel-Selke A, Steier C, Jurisch A, Tullius SG, Sawitzki B, Kolls J, Volk H-D, Ritter T: Targeting of macrophage activity by adenovirus-mediated intragraft overexpression of TNFRp55-Ig, IL-12p40, and vIL-10 ameliorates adenovirus-mediated chronic graft injury, whereas stimulation of macrophages by overexpression of IFN-g accelerates chronic graft injury in a rat renal allograft model. J Am Soc Nephrol 2003;14:214–225.

22 Lehmann TG, Wheeler MD, Schoonhoven R, Bunzendahl H, Samulski RJ, Thurman RG: Delivery of Cu/Zn-superoxide dismutase genes with a viral vector minimizes liver injury and improves survival after liver transplantation in the rat. Transplantation 2000;69:1051–1057.

23 Coito AJ, Buelow R, Shen XD, Amersi F, Moore C, Volk H-D, Busuttil RW, Kupiec-Weglinski JW: Heme oxygenase-1 gene transfer inhibits inducible nitric oxide synthase expression and protects genetically fat Zucker rat livers from ischemia-reperfusion injury. Transplantation 2002;74:96–102.

24 Ke B, Buelow R, Shen XD, Melinek J, Amersi F, Gao F, Ritter T, Volk H-D, Busuttil RW, Kupiec-Weglinski JW: Heme oxygenase 1 gene transfer prevents CD95/Fas ligand-mediated apoptosis and improves liver allograft survival via carbon monoxide signaling pathway. Hum Gene Ther 2002;13:1189–1199.

25 Braudeau C, Bouchet D, Tesson L, Iyer S, Rémy S, Buelow R, Anegon I, Chauveau C: Induction of long-term cardiac allograft survival by heme oxygenase-1 gene transfer. Gene Ther 2004;11: 701–710.

Ariela Benigni, PhD
Mario Negri Institute for Pharmacological Research
Via Gavazzeni, 11, IT–24125 Bergamo (Italy)
Tel. +39 035 319888, Fax +39 035 319331, E-Mail abenigni@marionegri.it

Author Index

Subject Index